Harriet Brooks: Pioneer Nuclear Scientist

Harriet Brooks

Pioneer Nuclear Scientist

MARELENE F.
RAYNER-CANHAM

GEOFFREY W.
RAYNER-CANHAM

McGill-Queen's University Press
Montreal & Kingston • London • Buffalo

ISBN 0-7735-0881-3

Legal deposit first quarter 1992
Bibliothèque nationale du Québec

Printed in Canada on acid-free paper

This book has been published with the help of a grant from the Canadian Federation for the Humanities, using funds provided by the Social Sciences and Humanities Research Council of Canada. Publication has also been supported by Canada Council through its block grant program.

Canadian Cataloguing in Publication Data
Rayner-Canham, Marlene F.
Harriet Brooks: pioneer nuclear scientist

Includes bibliography and index.
ISBN 0-7735-0881-3

1. Brooks, Harriet, 1876–1933. 2. Physicists–Canada–Biography. 3. Women physicists–Canada–Biography. I. Rayner-Canham, Geoffrey. II. Title.

QC16B77 R3 1992 539.7'092 C91-090627-0

This book was typeset by Typo Litho composition in 10/12 Palatino.

CONTENTS

PREFACE

Harriet Brooks was one of the leading women experimental physicists of her day and one of the first Canadian researchers in the field of radioactivity. Her research, first with Ernest Rutherford, then with J.J. Thomson, and later with Marie Curie, was of considerable importance in the early study of the phenomenon of radioactivity. Rutherford acknowledged her significant contributions in his two classic texts on radioactivity, and he commented that next to Curie she was the best woman scientist in the field.

The life of Harriet Brooks needs to be told – in part, to fill a missing piece of the history of research into radioactivity, but more importantly, to describe what it was really like for a woman to perform physics research during that period. We need to show that Marie Curie was not the only woman physicist of the period, and that of those others, Brooks was certainly among the most significant contributors to physics research in the early days.

Although we will focus on Brooks as a pioneer nuclear scientist, we believe it is important to convey a more complete biographical account of her life. In addition to her role as a physicist, her encounter with Maxim Gorky and her later married life in Montreal show the breadth of her interests and knowledge. Too often scientists are portrayed as one-dimensional creatures, and it would be a disservice to the memory of Harriet Brooks not to narrate the other aspects of her life.

We are fortunate in having a significant number of letters to call upon. Many of these are Brooks's own, while some are from physics colleagues and from social acquaintances. These letters give us a window into the past. In our time, such letters would simply be discarded, that is, if we had not used the telephone, electronic mail, or fax machine instead.

At the same time, we are disadvantaged by a lack of personal reminiscences by Brooks. In fact, it has been difficult to find commentaries by any women pioneers in physical science from that period. To give the reader an idea of the challenges she faced, we have had to resort to quoting the experiences of other women from slightly earlier or later time frames; indeed, as we point out, some of the difficulties she would have experienced persist even today.

Our lengthy toil on this biography would not have been possible without the enthusiastic support of many archivists. Their individual contributions are recognized in the appropriate locations in the text. However, special mention should go to Elizabeth Behrens, librarian, Sir Wilfred Grenfell College, whose help, particularly in the early stages, guided us into the uncharted waters of historical research. She also obtained many items of source material for us. Similarly, we thank the staff of the inter-library loan service at the University of California at Santa Cruz, who, during our sabbatical leave there in 1988–89, so patiently tracked down and ordered an enormous number of rare and obscure items. We are also grateful to Phebe Chartrand of the archives of McGill University for unearthing the academic records of Brooks and the correspondence of Principal Peterson.

As scientists, we found that writing an historical account was a voyage into new territory, and in this regard we are most grateful to James Greenlee, Department of History, Sir Wilfred Grenfell College, for a critical review of the first draft of this manuscript. Also, thanks are due to Marianne G. Ainley, Concordia University, for her constructive comments on the manuscript.

Finally, our biography would have been incomplete but for the help and encouragement of Paul Brooks Pitcher, son of Harriet Brooks, and of Cicely Grinling, daughter of Elizabeth Brooks, Harriet's sister. In the same acknowledgment, we are indebted to Margaret Gillett, McGill University, who supplied us with the telephone number that enabled us to contact the surviving members of Brooks's family.

The Brooks family portrait, probably taken in Seaforth in 1890, when Harriet was fourteen years old. From left to right, front row: Nelson, Herbert, May; middle row: George Brooks, Elizabeth (Worden) Brooks, Georgina; back row: Harriet, Elizabeth with James, and Edith. (Photo courtesy of P.B. Pitcher)

Graduation photograph of Harriet Brooks, 1898. (Notman Collection, McGill University Archives)

Members of the physics department, McGill University, taken on the steps of the Macdonald Physics Building (ca. 1899). Harriet Brooks is conspicuous at the rear. Ernest Rutherford is the figure furthest right. (Photo courtesy of P. Fowler)

The Macdonald Physics Building. (McGill University Archives)

Prestonia Martin and Harriet Brooks sitting by the fireplace at Summerbrook, 1906. (Photo courtesy of E. Jewitt)

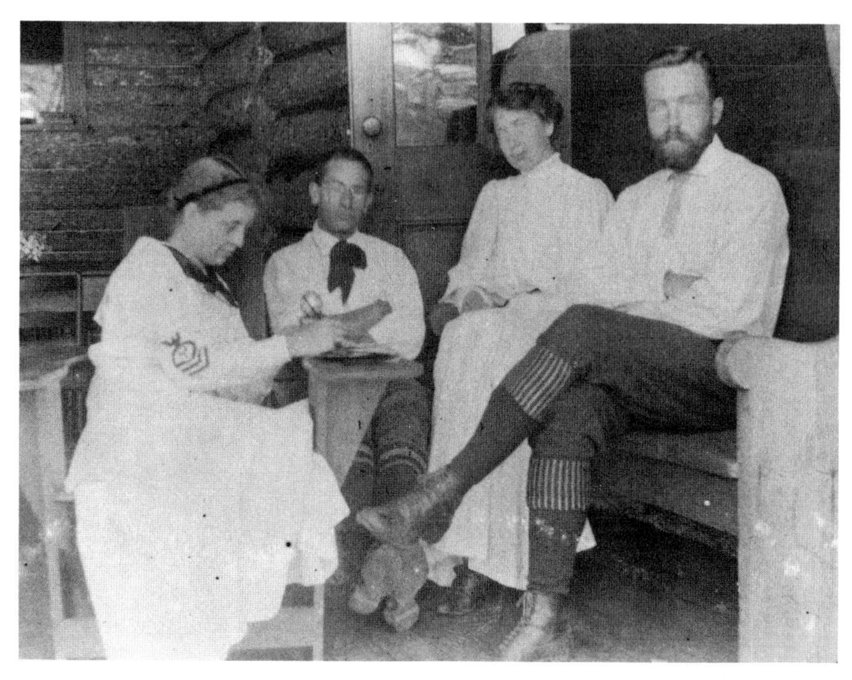

Brooks at Arisponet. From left to right, Prestonia Martin, Zinovy Peshkov, Brooks, and Nikolai Burenin. (Gorky Museum)

Brooks, Maria Andreyeva, and Maxim Gorky on the transantlantic ship *Prinzess Irene*, 1906. (Gorky Museum)

From left to right, Gorky, Burenin, Brooks, and two others on Capri, 1906 or 1907. (Gorky Museum)

From left to right, unknown woman, Gorky, Burenin, Andreyeva, I. Ladyzhnikov or L. Andreev, and Brooks on Capri, 1906 or 1907. (Photo courtesy of P.B. Pitcher)

Harriet with her three young children, Barbara, Charles, and Paul, ca. 1914. (Photo courtesy of P.B. Pitcher)

Harriet Brooks Pitcher and Frank Pitcher enjoying a picnic in July 1914, probably at the fishing camp on the St. Lawrence. (Photo courtesy of P.B. Pitcher)

Harriet Brooks, undated. (Photo courtesy of P.B. Pitcher)

Harriet with Paul and Barbara, ca. 1915–16. (Photo courtesy of P.B. Pitcher)

Harriet with Barbara, Charmaine Martin (daughter of John and Prestonia Martin), and Paul, ca. 1926. (Photo courtesy of P.B. Pitcher)

The Pitcher family (Barbara, Frank, Charles, Harriet, and Paul), ca. 1921. (Photo courtesy of P.B. Pitcher)

Harriet Brooks: Pioneer Nuclear Scientist

CHAPTER 1

HARRIET BROOKS: PIONEER WOMAN PHYSICIST

There is very little documentation on early Canadian women scientists,[1] and most of those whose names have come to light worked in the biological sciences. Thus, the "discovery" of a pioneer Canadian woman physicist fills a significant gap in our knowledge of the role of women in early Canadian science. In this book, we will look at the life and work of Harriet Brooks, whose contribution to nuclear science has been overlooked for almost ninety years, and we will discuss her story in the context of the avenues open to women during her era.

From the perspective of the late twentieth century, it is difficult to appreciate the challenges that faced Harriet Brooks in her pursuit of a career in physics. At the time of her youth, only two women had received acclaim for their work in physics. One of these was Laura Bassi (1711–78), who gained a chair in physics at the University of Bologna, Italy,[2] while the other was Émilie du Châtelet (1706–49), who is best known for her research on mechanics.[3] However, it is unlikely that Brooks even knew of their existence.

Brooks grew up during an era when the role of women in the sciences was changing. Until the 1880s, women tended to play a peripheral role,[4] but from then on there was a rapid increase in the involvement of women in mainstream science. As an illustration, in the United States only thirty-six scientific papers were written by women between 1874 and 1883, but this number increased to about four hundred in the period 1883 to 1900. This change was brought about by the increasing number of opportunities for women in higher education.

The pioneering institution in Canada was Mount Allison University, which opened its classes to women in 1862 and awarded the first Canadian degree granted to a woman in 1875.[5] Other universities followed, graduating their first women students in 1884 (Acadia and Queen's), 1885 (Dalhousie and Toronto), and 1888 (McGill).[6] Thus, it was only during Brooks's childhood that the option of a higher education for women became a reality. Even then, the opportunities were fewer than south of the border, where there were a number of independent colleges for women. In 1900, for example, only 11 percent of Canadian college students were women, while the figure was 36 percent in the United States.[7]

We are looking at a period in history when the idea of advanced education for women was still being challenged. The powerful arguments of the day have been summarized by Hubbard:

> More effective were the extensive treatises, replete with case histories, that "documented" the drain that menstruation and the maturation of the female reproductive system was said to put on woman's biology and, more importantly, the stress that would fall on these vital capacities if women's intellects were taxed by education. One of the most widely read books of this sort was Edward H. Clark's *Sex in Education*, published in 1873, which went through seventeen editions in the next thirteen years. Clark, a former professor at Harvard Medical School and a Fellow of the American Academy of Arts and Sciences, details the histories of many girls whose health, he assures us, was severely damaged by education.[8]

At the same time, there were some strong proponents of a university education for women, particularly a scientific education.[9] Three arguments were given in support of the proposal: that women would become better wives and mothers with a scientific background; that simple justice and equal opportunity should permit women to enter scientific careers; and that women had a greater capacity for noting details and superior patience and delicacy, enabling them to perform certain scientific work better than men. In fact, the case was made that women would benefit more than men from a training in science:

> Prevalent opinions and customs impose on women so much more monotonous and colourless lives, and deprive them of so much of the natural and healthy excitement enjoyed by the other sex in its free intercourse with the world ... Many women might be saved from the evil of the life of intellectual vacuity, to which their present position renders them so peculiarly liable, if they had a thorough training in some branch of science, and the opportunity of carrying it on as a serious pursuit.[10]

Brooks was an outstanding undergraduate student at McGill, but then so were many other women. The key to later success in her case was the offer of graduate work with Ernest Rutherford,[11] one of the leading researchers in the rapidly growing field of nuclear science. Rossiter notes that women were more likely to enter and be welcomed into fields that were rapidly growing (where there was a shortage of personnel) or were stagnant or shrinking (where men were less willing to endure the consequent hardships) than in fields of modest growth. She adds: "There is some feedback mechanism that channels women scientists into those fields most open or receptive to them. But the process by which women scientists choose fields ... remains to be explored."[12]

In our view, the role played by a supportive supervisor in encouraging women researchers has been overlooked. An example from a different field is that of X-ray crystallography, a subject pioneered by Sir William Bragg during the 1920s. Kathleen Lonsdale comments that it was the influence of Bragg at the Royal Institution that resulted in the number of women researchers there rising to 20 percent of the total.[13] Several of the women who worked with Bragg, including Lonsdale[14] and Dorothy Hodgkin,[15] attained world renown in crystallography.[16]

Like Bragg, but unlike many of his contemporaries,[17] Rutherford encouraged women students. Rutherford played a central role in Brooks's professional life; after she left McGill, it was to him she turned for advice, assistance, and inspiration. She was not the only woman to join his research group around that time, though she was certainly the most productive. From the names on published papers, we can identify three of these other women students: Fanny Cook Gates, who worked with him at McGill, and, after his move to Manchester, May Sybil Leslie and Jadwiga Szmidt.[18] Rutherford always acknowledged the contributions of women researchers, and in his two classic works on radioactivity, he cited work by many women scientists.[19] In the Bakerian Lecture in which he first described the idea of successive radioactive decay, he repeatedly acknowledged the work of Brooks (and mentioned that of Gates).[20]

Rutherford also had a reinforcing role with women researchers outside his own group. Ellen Gleditsch,[21] who worked with Marie Curie[22] and Bertram Boltwood,[23] wrote a letter to Rutherford in which she thanked him for his kindly conversations and encouragement at a conference in Washington in April 1915. Marie Curie, too, took time to write to Rutherford and express her thanks for "all his little kindnesses."[24]

The feminist historian Helena Pycior interprets a comment Rutherford once made as indicating that Rutherford did not encourage women scientists.[25] Rutherford had remarked to Lise Meitner upon first meeting her that he had thought she was a man:[26] Pycior thought the remark suggested that Rutherford did not expect that a woman would have performed such research. However, we prefer a more innocent explanation. In Rutherford's two texts, the indexed names of women researchers are prefixed by "Miss," "Mrs," "Mlle," "Mme," or "Lady," as appropriate. However, since Meitner is simply listed as "L. Meitner," Rutherford may not have realized that she was female.

Comments by the astronomer Cecilia Payne-Gaposchkin have also been used to cast an unfavourable light on Rutherford's attitude towards women.[27] However, Payne-Gaposchkin's specific evidence for Rutherford's "scorn" towards women was his habit of starting his lectures with the phrase "*Ladies* and Gentlemen." Although this emphasis on "ladies" must have been intimidating to as shy a person as young Payne-Gaposchkin, we strongly believe that the letters from Rutherford's women students and correspondents indicate a supportive disposition.

Nuclear science seems to have been an acceptable branch of science for women. It is amazing how many women were involved in the study of radioactivity during that period. As well as Brooks, Gates, Gleditsch, Leslie, and Szmidt, there were at least ten other women who authored or co-authored papers on the subject [28] – a significant proportion of the researchers of the time. As well, there appear to be several others whose work did not result in publication.

When we look at Brooks's career after graduation, we see the problem of employment for a woman graduate. Antler has shown that most single women graduates actually returned to the parental home, some as "working daughters," while others became "ladies of leisure."[29] Few left home to pursue their own ambitions (the "independents") as Brooks did.

Rossiter points out that when opportunities for higher education for women increased during the latter part of the nineteenth century, little thought was given to the eventual careers that such graduates might take up.[30] She has identified three directions of employment: the "big science" areas where demand for personnel resulted in the hiring of qualified women support staff (such as astronomy);[31] women's colleges and (US) co-educational land-grant agricultural colleges, where there was a need for new faculty and other personnel; and those fields that became known as the "female sciences," such as home economics. Brooks's research work would fall into the first category, while her subsequent teaching at Barnard College would fit into the second. For Canadian women, college-teaching opportunities were so few in Canada that emigration to the United States became almost a necessity.

Women found it very difficult to break into the ranks of professorship at a "normal" institution, as Talbot noted back in 1910: "Except in some of the women's colleges where the opportunities for research are limited and the salaries notably low, women are not considered eligible for chairs in the sciences. If they have any positions in the departments at all, it is chiefly as laboratory assistants."[32] In other words, laboratory directors would hire the best researchers, regardless of sex, but the upper leadership ranks were still considered to be male preserves.[33] Later evidence of this comes from the experience of Jane M. Dewey, who in 1929 was unable to find a post at a regular university, in spite of her excellent credentials.[34]

In her decision whether or not to marry, Brooks must have been aware of the low marriage incidence among academic women. Kendall summarizes the situation as it was at about the turn of the century:

In 1895 the entire women's college movement was shaken by the announcement that fully half the graduates had remained spinsters ... The spinster explosion was taken very seriously by the American public, and the colleges were naturally held responsible for it ... One misguided attempt to justify the colleges implied that educated women were no great loss to the race since they were "more or less lacking in normal sex instincts" and that the college probably served a useful purpose by segregating such women from the rest of society.[35]

A similar situation existed in Canada, where about half of the 392 women who studied at Dalhousie University between 1885 and 1900 remained single as compared with 10 percent unmarried among the total female population.[36]

A popular and more positive argument than that suggested by Kendall was that women who excelled academically should regard it an honour and a duty to devote their lives to knowledge to the exclusion of a family life: "Civilization rests upon dedicated lives, lives which acknowledge obligation not to themselves or to other single persons, but to the community, to science, to art, to a cause."[37]

Yet there was no equivalent requirement for men to be single. In fact, one must keep in mind that most of the male physicists, including Rutherford and J.J. Thomson,[38] were married. Fox Keller points out that the academic life of a single woman may not be as productive as that of a married man who has a wife at home to take care of the family's affairs.[39] The suggestion is that even the single woman may carry a double burden.

Like Brooks, many other well-educated women of the day were faced with the alternatives of marriage or a career.[40] A quote from the diary of a Vassar graduate of 1902 describes the cross-pressures confronting educated women:

So much of the trouble is because I am a woman. To me it seems a very terrible thing to be a woman. There is one crown which perhaps is worth it all – a great love, a quiet home, and children. We all know that is all that is worth while, and yet we must peg away, showing off our wares on the market if we have the money, or manufacturing careers for ourselves if we haven't. We have not the motive to prepare ourselves for a "life-work" of teaching, of social work – we know that we would lay it down with hallelujah in the height of our success, to make a home for the right man.

And all the time in the background of our consciousness rings the warning that perhaps the right man will never come along. A great love is given to very few. Perhaps this make-shift time filler of a job *is* our life work after all.[41]

We will see Brooks's story as one typifying the challenges facing a woman scientist of the period. She entered a new field of science and, with the encouragement of Rutherford, performed valuable research. Ultimately, she was faced with having to choose between a career with limited opportunities for women and the security of marriage and a family, an either/or choice that still exists to a certain extent today.

CHAPTER 2

THE BROOKS FAMILY

Harriet Brooks was born on 2 July 1876 in Exeter, Ontario, a small town in western Ontario about forty-eight kilometres from London. The area was first settled in about 1832 and the village itself was founded in 1851. Many of the early pioneers had come from the county of Devon in England, and they named the town after Exeter, the county seat.[1] A number of industries were established in those early years of settlement: a tannery, a furniture factory, and then, later on, flax, flour, grist, saw, and woollen mills. Since Exeter was the chief market centre of the district, its population had grown to about 1,000 by 1873.

Harriet was the daughter of George Brooks of Mitchell, Ontario, and Elizabeth Agnes Worden of Exeter.[2] George Brooks's father came from Londonderry, Ireland, and is thought to have been a weaver by trade. There is no earlier information on the Brooks side of the family, but we can trace the Wordens back to their arrival in Canada. John Worden and Susan (Tapson) Worden moved to Canada in 1836, about a year after their marriage at Bridestow, Devonshire, England. The Wordens were among the early waves of settlers from Devon. During the mid-nineteenth century, there was a massive emigration from that county as a result of the low wages paid to farm workers. By the end of the century, half of the rural population had left Devon in search of a better future.[3] In proportion, this was a greater migration than that from southern Ireland after the famine.

The Wordens settled initially in Bannockburn in eastern Ontario, a small settlement with a grist mill,[4] then moved to Exeter sometime between 1855 and 1870. The Wordens had eight children; Elizabeth Worden was born on 29 April 1850.

George Brooks and Elizabeth Worden were married on 14 July 1871 at Osborne Manse, Exeter, by the Rev. Henry Gracey. They had nine children: Georgina Susan (5 July 1872); Edith Annie (4 June 1874); Harriet (2 July 1876); Elizabeth Agnes (13 August 1878); Herbert Franklin (4 November 1880); George Nelson (13 November 1882); Lydia May (21 June 1885); Isabella Kirkton (14 April 1887); and James Ernest (13 June 1889). Isabella died of diphtheria at the age of seven months, while Lydia died of peritonitis at thirteen years.

Harriet was the only child not to have a middle name listed in the family Bible, nor did she ever use one herself. However, in her joint publications with Ernest Rutherford she is listed as Miss H.T. Brooks. The "T" might have represented the grandmother's family name of Tapson.[5]

The Brooks family initially lived in Forest, Ontario, about forty-three kilometres from Exeter. Forest, with a population of nearly 2,000 at that period,[6] was the centre of a thriving farming region, possessing a number of agricultural-related industries. George Brooks was the owner of a flour mill in Forest that burnt down uninsured. A ball of metal with the grains of wheat embedded in it, all that was left of the mill, was subsequently kept on the Brookses' mantelpiece. The family then moved to Exeter, and the father took a job as a flour commercial traveller with W.W. Ogilvie, a position he kept the rest of his life. According to Brooks family history, George Brooks was said to "comport himself like a single gentleman," and this may be a reflection of his life on the road.

The Brooks family moved to St. Dominique Street, Montreal. They stayed there for three years and then returned to Ontario, this time to Seaforth in Huron County in 1888, where they lived next to a rectory and church. The town of Seaforth was the result of a business venture, a means of luring the Buffalo and Lake

Huron Railway from neighbouring communities.[7] The subsequent arrival of the railroad assured prosperity for the town, and it developed into a busy grain market, as farmers from far away brought their produce, mainly wheat, to town for shipment by rail. Seaforth had three flour mills, one of these being the Ogilvie & Company mill.

It involved considerable hardship to feed and clothe eight children on the salary of a commercial traveller. Elizabeth Brooks often recalled how visitors were always welcome to a meal, but offering such hospitality wasn't always easy. Family members became familiar with the two codes: FHB (family hold back) if there was a shortage of food, and MIK (more in kitchen) if there was plenty to go around. Elizabeth remembered that she never had a new dress until she earned money herself from teaching.

Both Elizabeth and Harriet went to school in Seaforth. The town had had a high school since 1879, and in the year the Brooks family arrived, 1888, the school had been elevated to the status of a collegiate institute.[8] The Seaforth Collegiate Institute possesses the school records back to 1893; that year shows an entry on the register for "Lizzie Brooks,"[9] Harriet's younger sister, Elizabeth. There is no record of Harriet, since at age seventeen in 1893 she would have already graduated.

Presumably, Harriet acquired her interest and basic skills in mathematics and the sciences during her high school education. It would be fascinating to know what influenced her to study (and excel in) such non-traditional subjects for women. At the time, and even to the present day, female children are not raised in an environment that encourages the development of math and science skills. This point was noted even as early as 1910:

> Whoever will watch groups of girls and boys in any grade school must realize that out of sight, in the homes, distinctions are introduced which result ultimately in mental handicap for the girl.
>
> The proprieties and dainty clothing cost her many a lesson that her brother learns; and who concern themselves to take a girl to the blacksmith shop, the power-house, and the stone-quarry, to the places where the steam-shovel and the pile-driver are at work.
>
> Given the same circumstances, including the circumstances of encouragement, and it is hardly to be doubted that the rational curiosity

to know the causes of things would be found in girls as it is in boys. Opportunity is rendered ineffective and the world of natural phenomena inviting to observation and analysis is denied to girls because they are assigned to an artificial environment demanding an emotional response; and then we wonder at it when young women in their junior and senior years at college elect music and literature in preference to mechanics and physiology; we wonder and we frame theories about feminine predilections.[10]

The family moved again, first to Sherbrooke in the Eastern Townships of Quebec, and then back to Montreal to an upper two-floor tenement up long outside stairs on St Denis Street, further east than before but in a healthier, wider street. By this time, Harriet had been accepted into McGill University.

Harriet and Elizabeth were the only two of the family to go to university. What spurred the Brooks sisters to continue their education? Not enough is known about the family life to give an exact answer, but Elizabeth Brooks' daughter, Cicely Grinling; sheds some light on the matter:

I really think my Grandmother Brooks was the strong, possibly broad-minded, character in the family and that her husband was weak, wayward and opted out of all responsibility after his mill burnt down and found solace in travelling as a salesman and spending considerable time drinking in saloons.

I always had the impression from my mother that it was the lack of opportunity and parental encouragement that spurred her [and her sister] on to academical prowess. Books were rare and entailed long walks to public libraries – so that it was the challenge which encouraged her.[11]

Throughout her life, Harriet Brooks was particularly close to her sisters, Elizabeth and Edith. Elizabeth was equally gifted and she obtained a two-year scholarship to McGill University in 1896. It was a classics scholarship, but the final two years of classics were not open to women at that time. As a result, Elizabeth taught herself enough mathematics to get a mathematics scholarship, and also achieved an honours classics degree when the courses were later opened to women. She became a schoolmistress and, at the age of twenty-five, the headmistress of a boarding school.

She married Arthur Stewart Eve in 1904.[12] Eve was a physicist at McGill University who rose to the rank of professor of physics and then dean of graduate studies.

Edith Anne Brooks became secretary to Charles Blair Gordon, a prominent Montreal businessman, subsequently marrying him in 1897.[13] Gordon had worked his way up through the Dominion Textile Company to become the president in 1909. He was also elected to the boards of directors of a number of major Canadian companies and became president and chairman of the Bank of Montreal. Gordon was knighted for his services to the British Purchasing Commission in Washington during the First World War.

Of the other siblings, Georgina Susan Brooks went to Europe at about twenty-three years of age and never returned, living alone until her death in her eighties. The family of George Nelson Brooks lived next to the Eves in Montreal. James Ernest Brooks (died 1945) married a "Southern Belle." Herbert Franklin Brooks (died 1939) emigrated to the United States.

CHAPTER 3

DEFYING CONVENTION: THE McGILL YEARS

It must have been an extraordinarily courageous decision of Harriet Brooks's to decide to go to university. It had been barely six years earlier, in 1888, that the first woman had graduated from McGill and only twelve years since the first woman had obtained a university degree in Canada.[1] Many people, clinging to Victorian stereotypes, still believed that women should not be permitted to participate in higher education. This view was reinforced by prominent academics, such as the sociologist Herbert Spencer, who concluded that the differences between the sexes could best be understood in terms of "a somewhat earlier-arrest of individual evolution in women than in men."[2]

Solomon comments on the women who went to American colleges at this time:

> The early females who contended for places at collegiate institutions wanted that education passionately. But they needed encouragement at a time when their objective defied convention. Nothing was more crucial among the factors accelerating women's advance into academic institutions than family support. Ordinarily the aspiring young woman had to take the initiative with the family in negotiating her educational future: her parents were unlikely to pressure her to continue her schooling. The decision was a major one in most families. Assent depended on available financial resources and recognition of the value of a college education for the particular daughter.[3]

The poorest families were simply unable to afford to send their children to university, while the rich would choose an education that would prepare their daughters for a life of leisure rather than one of intellectual activity.[4] The majority of the women students came from the expanding middle class, from professional and business families.

Although parental support was important, the first generation of female students at Dalhousie University (1881–1900) relied more on self-motivation.[5] Such a strong ambition was probably the driving force behind Harriet Brooks's success. Her feelings may have been similar to those of M. Carey Thomas, who graduated from Cornell about nineteen years before Brooks entered university: "The passionate desire of the women of my generation for higher education was accompanied through its course by the awful doubt, felt by the women themselves as well as the men, as to whether women as a sex were physically and mentally fit for it ... I was always wondering whether it could be really true, as everyone thought, that boys were cleverer than girls."[6]

Vicinus notes that going to college was one of the few ways a daughter could escape from the parental home.[7] To break away required a strong self-image, particularly if the family held to the conventional view that a well-brought-up Victorian and Edwardian girl should stay quietly at home until a suitor appeared on the horizon. As an incentive, a college education leading to a teaching position (a higher-paying position than most others available to women) offered a middle-class girl the reward of financial independence.

The education of women at McGill came about as a result of financial support from the railway tycoon Sir Donald Smith (later Lord Strathcona).[8] In recognition of his role, women students were listed as members of the Donalda Department and were commonly referred to as Donaldas. Probably as a consequence of the strong beliefs of the principal, Sir William Dawson, Smith's gift required that the sexes be educated separately. The range of courses available to women was more limited than those for men, for the lecturers had to repeat their performance for the benefit of the women's class and not all were willing to do so. As well as these repeat performances, additional tutors had to be hired to instruct the women. The women were taught in separate classrooms with separate waiting rooms. They even had separate entrances and stairways.

Most of the third- and fourth-year courses were co-educational, as it would have been impossible to justify the cost of teaching the one or two women students in these specialized courses separately. The women's examinations were the same as those for men, and women had the same privileges in terms of classing, honours, prizes, and medals.

Even at the university the societal pressures on women students were not relieved. As was remarked by one of Brooks's predecessors, each of the Donaldas was conscious that "she bore the weight of formulated womanhood upon her shoulders, although men, even then, were not expected to live to the ideal man."[9] A young woman would have also needed strength of character to face the stares and comments of the men students. For example, Elizabeth Irwin recalled that "it required courage in those days to walk from the East Wing to Molson Hall, or the old Library below the Hall. It meant ... running the gauntlet of the men students, who, not yet accustomed to the intrusion of the feminine element, greeted our appearance with the strains, long since forgotten, of 'Hop Along, Sister Mary.'"[10]

The elaborate clothing of the time put women at an additional disadvantage in the study of science. A physics demonstrator of this period at the Royal College of Science, London, writes: "When I was a demonstrator I had some young women in my group and when they were working with their suspended magnet galvanometer the needle wandered about as they moved. They complained to me of this and shyly I had to tell them that it was the 'steels in their stays' that caused the trouble."[11]

Impediments notwithstanding, Harriet Brooks performed outstandingly well throughout her undergraduate education. To give an idea of the education of the time, the courses that she took and her ranking are listed below:[12]

First Year
Ancient History (I); Chemistry (II); English (I); French (I); Geometry and Arithmetic (I); German (I); Greek (III); Latin (II); Trigonometry and Algebra (I)

Second Year
French (I); Geometry and Arithmetic (I); German (I); Greek (I); Latin (II); Latin Prose and Composition (I); Mental and Moral Philosophy (I); Modern History (I); Trigonometry and Algebra (I)

Third Year
Astronomy and Optics (I); Experimental Physics (I); French (II); Laboratory Course (I); Mechanics and Hydraulics (I)

Fourth Year
Experimental Physics (I); Laboratory Course (I)

In each year of her studies Brooks obtained first-rank general standing and first-rank honours. Her success in the first year resulted in the prize in mathematics and the award of a second-year exhibition (scholarship), valued at $100 plus free tuition, donated by Sir Donald Smith. In the second year, Brooks obtained the prize in german and, for her excellence in mathematics, she was awarded a mathematical scholarship, tenable for two years and valued at $125 per year, donated by Lord Strathcona. With George Brooks's limited income and the large number of mouths to feed, it must have been a financial struggle to send Harriet and Elizabeth to university. Harriet's scholarships would have been of great help.

Brooks was a popular student and in her third year was elected class president.[13] In 1898 she graduated with first-rank honours in an honours degree in mathematics and natural philosophy (the only student to do so, that year). In addition, she obtained the highly coveted Anne Molson Gold Medal for outstanding performance in mathematics. She simultaneously obtained a teaching diploma.[14] This was not uncommon for McGill's women students at the time, as it opened the option of a teaching career, one of the few acceptable professions for an educated woman. The diploma was awarded by the McGill Normal School, a teacher training school associated with McGill University. To fulfil the requirements of the diploma, the McGill students took pedagogy courses in the evenings. Tuition at the Normal School was free, but in return students had to sign a contract committing themselves to three years of teaching.[15]

After her graduation, Brooks was invited to join the research group of Ernest Rutherford. Rutherford had just arrived from the Cavendish Laboratory at Cambridge University, having been lured to McGill by the new Macdonald Physics Building that had been opened in 1893.[16] The funds for this and the engineering building were supplied by the Montreal tobacco merchant Sir William Macdonald, who, after attending a meeting of the British Association for the Advancement of Science, became increasingly interested in science.[17] The building itself was described as the best physical laboratory in North America, while its collection of physics apparatus was said to be the most comprehensive in the world.

For a woman to enter graduate work in physics was a significant breakthrough. Even in the 1970s, it was commented that "to get as far as she has, a woman physicist must probably be better than the men physicists she meets along the way. A girl has to have an unusually strong motivation to become a physicist. And she encounters prejudice all along the road."[18]

Initially, Brooks worked in the field of electricity and magnetism. Her research concerned the decrease in magnetization of steel needles when an electrical discharge is passed through the needle. It would seem surprising that Rutherford would set her to work on this topic rather than on some of the radioactivity experiments for which he was already becoming famous. The magnetization and demagnetization of steel needles had been Rutherford's area of research when he did his undergraduate thesis at Canterbury College, New Zealand, and during his first years at the Cavendish Laboratory.[19]

It is possible that at this time Rutherford lacked the radioactive materials necessary to start his students off with the research into the nature of radioactivity. The evidence comes from Feather, one of Rutherford's biographers, who noted: "A day or two before he left England, also, Rutherford ordered, through the secretary at the Cavendish Laboratory, some uranium and thorium salts with which to continue his work on radioactivity. On 24th October, 1898, he wrote reminding the secretary of this order and asking that the material should be forwarded as soon as it arrived."[20]

An alternative explanation is that because radioactivity was a largely unexplored area, Rutherford thought it wiser to start

Brooks on a more established topic. In his recollections of Rutherford, Kapitza comments: "He was also very particular not to give a beginner technically difficult research work. He reckoned that, even if a man [or woman] was able, he needed some success to begin with. Otherwise he might be disappointed in his abilities, which could be disastrous for his future. Any success of a young research worker must be duly appreciated and must be acknowledged."[21]

In any event, Brooks's research went well. Her results were presented by a Mr Deville (as she was not a member) to the Royal Society of Canada on 26 May 1899 and were subsequently published in the *Royal Society of Canada (Transactions)*.[22] Not until 1901 was the work submitted as a thesis towards a master's degree. The Faculty of Arts approved the thesis on 25 January 1901[23] and on 29 March 1901 gave leave for the contents of the thesis to be published.[24] This was the first masters degree in physics awarded to a woman at McGill. At that time, an MA was the limit to which one could proceed in physics at McGill, the first physics doctorate not awarded until 1909.[25]

Brooks was Rutherford's first graduate student, although he had hired Robert K. McClung (BA 1899), possibly in 1898, when McClung was still an undergraduate student. By September 1899 Rutherford had four students working with him, the new additions being A.G. Grier and S.J. Allen.[26]

In the fall of 1899, Brooks took up a position at the Royal Victoria College (RVC), while continuing her research with Rutherford.[27] The Royal Victoria College represented the culmination of the plans for a separate women's college at McGill.[28] The college was administered by Hilda Oakley, the warden, who reported to the new principal of the university, Sir William Peterson. Initially, RVC had three staff tutors: two resident and one non-resident,[29] Brooks being the non-resident tutor (presumably she lived with her family).

The formal opening of the college, which took place in November 1900, was a grand spectacle for all of Montreal. We know that Brooks was a participant, for her name was included in the list of invited guests published in the Montreal *Gazette*.[30] Oakley recalled the throng of people who attended, including "a great concourse of students, men and women."[31]

Meanwhile, Brooks had started her own research in the field of radioactivity. It is important to realize that she was a true pioneer, for at the turn of the century very little was known about the phenomenon. The release of rays from uranium salts that darkened photographic plates had been first noted in 1896.[32] Researchers during the 1896–1905 period were trying to understand these strange observations, which could not be fitted into the contemporary framework of classical physics.[33]

Rutherford reported in January 1900 that thorium gave out some radioactive substance apart from the common radioactive rays.[34] This radioactive substance was most unusual in that it could be carried away by air currents. Rutherford gave the name of "emanation" to this puzzling substance, and Brooks's task was to study this emanation and determine its nature. It was not clear whether the emanation was a radioactive gas, a vapour, or a very finely divided powder.

Using a diffusion method, Brooks identified the emanation as a radioactive gas of molecular weight in the 40 to 100 range.[35] The results of the experiments were read to the Royal Society of Canada as a joint paper by Rutherford and Brooks entitled "The New Gas from Radium." This study was subsequently published in the transactions of the society.[36] Rutherford published a more general account of the same work in the journal *Nature* under his name alone, with the acknowledgment "In these experiments I have been assisted by Miss H.T. Brooks."[37]

This discovery was crucial in the progress of radioactive research. At that time, the radioactive elements were believed to maintain their identity while the radiation was being released. Identifying the gas as having a significantly lower molecular weight than thorium indicated that the emanation could not simply be some gaseous form of thorium.[38] It was this result that led Rutherford, together with Frederick Soddy, to the realization that a transmutation of one element to another had occurred.[39] The discovery of nuclear transmutation was a key step in modern nuclear science, and the contribution of Brooks's pioneering experiment has been long overlooked.

Ernest Rutherford was a major figure in the life of Harriet Brooks. He had come from rural New Zealand and throughout his life maintained his rough-hewn character. His friend Chaim

Weizmann describes Rutherford as "youthful, energetic, boisterous, he suggested anything but the scientist ... He was a kindly person but he did not suffer fools gladly ... Any worker who came to him and did not prove to be a first class man was out in short order ... With all this, Rutherford was modest, simple and enormously good-natured."[40] Thus, Brooks's long association with Rutherford indicates that he must have considered her to have excellent research skills.

Rutherford's kind nature is also remarked upon by his biographer, D. Wilson: "It will inevitably be hidden and forgotten that he was a man of exceptional personal kindness. Everyone who remembers Rutherford remembers this – that he was personally kind to them far and away beyond the normal behaviour of a pleasant human being."[41] At the same time, Rutherford did have his "bad days." Robinson comments:

> Newcomers learnt that the sight of Rutherford singing lustily "Onward, Christian soldiers" (recognizable chiefly by the words) as he walked round the corridors was an indication that all was going well ... his other habit of intoning a melancholy dirge (never completely identified, for obvious reasons) when work was not going well, or when he had found someone maltreating a treasured piece of apparatus ... Crimes against apparatus he kept in a special category; for such crimes he had little forgiveness and an uncomfortably long memory.[42]

Happily for Brooks, Rutherford was extremely supportive of women in science.[43] He treated them as professional equals, and he was quite forthright in his views in this regard. Many years later, in 1920, there was a furor at Cambridge University over whether women should be admitted to the university with the same privileges as men. Rutherford and his chemistry colleague, William Pope, wrote a lengthy letter to the *Times* of London, in which they state:

> For our part, we welcome the presence of women in our laboratories on the ground that residence in this University is intended to fit the rising generation to take its proper place in the outside world, where, to an ever increasing extent, men and women are being called upon to work harmoniously side by side in every department of human affairs. For

better or for worse, women are often endowed with such a degree of intelligence as enables them to contribute substantially to progress in the various branches of learning; at the present stage in the world's affairs we can afford less than ever to neglect the training and cultivation of all the young intelligence available. For this reason, no less than for those of elementary justice and of expediency, we consider that women should be admitted to degrees and to representation in our University, and should be invited to assist in maintaining Cambridge in close contact with every aspect of human affairs.[44]

Brooks continued working with Rutherford until 1901, her later work involving a comparison of the radiations from the elements thorium, uranium, radium, and polonium.[45] At that time, very little was known about the radiation given off by radioactive elements. In the twenty-three-page article summarizing their work, Brooks and Rutherford reported two separate experiments. Uranium had been shown to produce two types of radiation: α rays, which we now know are the nuclei of helium atoms, and the more penetrating β rays, which are fast-moving electrons. In the first experiment, a layer of uranium oxide was covered with aluminum foil, which absorbed the α rays. A magnetic field was then applied to the β rays. By altering the conditions of the experiment, Brooks and Rutherford deduced that the β radiation was deviated by the magnetic field. They concluded that the β radiation consisted of negatively charged particles moving with high velocities. This was a crucial discovery. In a comparison experiment, they showed that the β radiation produced by radium was deviated to the same extent. Thus, β rays were the same, irrespective of their source.

In the second experiment, Brooks and Rutherford looked at the degree of absorption of radiation by different substances. This was an extremely thorough investigation, since it explored the different radiations produced by the elements uranium, thorium, polonium, and radium. The decrease in radioactivity with time was plotted, and the curve was fitted to a complex exponential equation. As Rutherford later noted, the difficulty in that era was that researchers did not realize the complexity of the phenomenon.[46] Each of the four radioactive elements did not undergo just one transformation; instead, as we are now aware, a series of changes take place, each new element producing its own radiation. It was not until the period 1904 to 1910 that any sense was made of the series of transformations.

The apparatus that Rutherford and Brooks used for these experiments still survives. A.S. Eve, a later collaborator of Rutherford's, saved many items from the research laboratory, a difficult endeavour, as it was customary to cannibalize old equipment for inclusion in parts of later experiments. These pieces of apparatus were subsequently sorted and identified by F.R. Terroux, who published details of them, complete with photographs.[47] The equipment has since been included in the Rutherford Collection at McGill University.[48]

In early 1901, Brooks applied for a fellowship at Bryn Mawr, a women's college in Pennsylvania. It is likely that Rutherford encouraged her in this so that she might broaden her experience, and she may have planned to start towards a doctorate. She had already amassed enough experimental results, if not the required coursework, for a PhD. Principal Peterson of McGill University wrote a glowing recommendation for her on 22 March 1901:

> Since her graduation she has been continuously engaged in research work in the Physics Laboratory. In addition to her high standing as a scientific worker, Miss Brooks has given proof of other qualifications in the capacity of Tutor in Mathematics in the newly instituted Royal Victoria College for Women. I speak from personal knowledge when I say that her work among our women students is very highly appreciated, and that she possesses various qualities which go to make a successful teacher, as well as those which will always make her an acceptable inmate of a residential College.[49]

Brooks was successful in obtaining the fellowship, and she moved to Bryn Mawr in 1901.

CHAPTER 4

A YEAR AT BRYN MAWR

Opened in 1885, Bryn Mawr was a women's college. However, it looked more towards Johns Hopkins University as a model than to other women's colleges, priding itself on standards, curriculum, and scholarship.[1] The institution was strongly dominated by its first president, M. Carey Thomas. Thomas tried to educate the students to be independent, competitive women who would never succumb to passivity. The Bryn Mawr women were seen by outsiders as eccentric, elitist, unusually intelligent, and markedly undomesticated. The graduates were not prepared for marriage but for independent careers, particularly in academia, in which they were expected to excel. As Wein comments: "The serious, scholarly atmosphere spread throughout Bryn Mawr's stark, Gothic buildings. After attending morning chapel and hearing an incisive talk by Carey Thomas, usually exalting the intellectual life, students paraded to class in caps and gowns. The faculty taught their courses in academic robes. Examinations were frequent and often traumatic. Grades were posted so that students felt enormous pressure to avoid embarrassment and succeed in their studies."[2]

Thomas had always argued particularly strongly for the need for graduate schools for women. She felt that for a teaching career, specialist knowledge above that required for an AB was needed. She also believed that it was essential that the faculty conduct research and teach advanced-topics courses. In her view, this was the only way to prevent faculty from going stale. Primarily, however, she believed that a graduate school would nurture brilliant women students and provide an atmosphere where they could be encouraged to progress to their full potential. Thomas argued: "If the graduate schools of women's colleges could develop one single woman of Galton's 'X' type – say a Madame Curie, or a Madame Kovalewsky[3] born under a happier star – they would have done more for human advancement than if they had turned out thousands of ordinary college graduates."[4]

Bryn Mawr had set up a graduate school from its very inception. Fellows, supported by grants from the college and selected from the graduates of other institutions, were to be the source of the graduate students, at least at first.[5] Brooks was one of these students, and she took the appropriate courses towards a PhD in physics. At that time, Bryn Mawr possessed the fourth-largest graduate school for women in the United States, behind Columbia, Chicago, and California. There were usually between sixty and seventy graduate students, a significant proportion of whom were from Britain and Canada.[6]

Brooks took a combination of post-majors courses and graduate courses. In the first semester, she took the following: Mathematics – Analysis, which dealt with determinants, Fourier's series, infinite series, definite integrals, and so on; Applied Mathematics – Thermodynamics, which covered the newer applications of thermodynamics, especially the work of Willard Gibbs, Helmholtz, and van't Hoff; Physics – Sound in Relation to Music, which began with a detailed mathematical discussion of the propagation of a sound wave and then introduced the work of Helmholtz and König, emphasizing its bearing on music; and Physics – Optics, which focused on the theory of Maxwell and the commentaries upon it.

In the second semester, Brooks took the second halves of the two-semester mathematics analysis and optics courses. In addition, she took an applied mathematics course and two physics courses: Physics – Spectrum Analysis, a study of the methods of spectrum analysis and of the distribution of spectrum lines; and Physics – Laboratory and Research Work. The laboratory work was very demanding, being listed as twenty-five hours per week, though the catalogue entry notes that "students taking physics as their chief subject for the degree of Doctor of Philosophy are expected to spend all the time possible in work in the laboratory."[7]

The building used by the science students was certainly well equipped, having special rooms for magnetic, optical, and electrical work, and a constant temperature vault in the basement designed for accurate comparison of lengths. The pedagogy was quite interesting, as the research work involved the progression from replicating previous experiments, to performing extensions of these known experiments, to performing new and original research.

During her career, Brooks continued to write periodically to Rutherford. Fortunately many of these letters survive as part of the Rutherford Collection of Correspondence,[8] and five of these pertain to the period of her stay at Bryn Mawr.[9] The address given by Brooks is that of the East Wing of the newest and largest residence building of Bryn Mawr, Pembroke Hall, completed in 1894.[10] The first letter describes how she had to build all her own research equipment from old laboratory apparatus. Her account of the research she was performing would indicate that she started research activities in the first semester, rather than the second as her program would indicate. She informs Rutherford that she will be home for the Christmas holidays, at which time she will tell him about her work so far. She closes the letter with this comment: "Please do not give me any more credit than I deserve in that comparison of radiations. You are quite too generous in that respect."[11]

It is very noticeable that Brooks minimizes her own contribution and abilities in letter after letter, yet the comments of Rutherford and others indicate that she was a very skilled physics researcher. According to Widnall, this attitude is a common phenomenon among female scientists.[12] In particular, women tend to suffer a significant loss of self-esteem in the sophomore year of college, while men maintain the same level of self-esteem. As a result, women arrive at graduate school with some uncertainty about their abilities even though their academic records and test scores are equivalent to those of the men.

In the next letter, while Brooks again belittles her research work, she is ecstatic at the honour awarded to her at Bryn Mawr:

I received the copies of the C.R.S paper which you sent me last week.[13] I'm afraid that your generosity in placing me as a collaborator where I am really nothing more than a humble assistant has rather imposed on the faculty of Bryn Mawr, for last night, they awarded me the European Fellowship and much to my satisfaction would prefer that I should make use of it at Cambridge. Of course I am just longing to be able to take advantage of it, the great difficulty is that I am afraid it may lose me my place at McGill and my experience down here has convinced me that there is no place where I want so much to be. It leaves so much to ask however that my place should be kept for me another year. Do you think it would be possible that they would get a substitute and let me come there the year after. President Thomas says that if I cannot use it this coming year, they will hold it over for me until the year after but I should like so much to be able to go as soon as possible.

I feel rather guilty about taking it for I certainly have not done anything worth mentioning since I came here, but they all seem very anxious that I should take it. For that I think I am greatly indebted to you for your good reports of me when you were here. I cannot thank you enough. There are of course other difficulties to be met besides the ones I have mentioned money and the objections of my family for instance who will think me wholly out of my mind but I think I can overcome them.

Will you put in a good word for me with Dr. Peterson I am going to write to him about it. Perhaps it would be better for me to come back for a year if they would give me a good salary so that I could have some money to go on. What do you think about it?[14]

There were three President's European fellowships awarded each year. The fellowships were designed to enable each recepient to experience a year at one of the leading European universities. One of the awards was given to the most able graduate student of one year's standing, and this would have been the category under which Brooks was considered. Flexner remarks that many of the European fellows return to Bryn Mawr after their year abroad to complete their training and receive the degree of doctor of philosophy.[15]

The financial problems that Brooks mentions again reflect that, unlike many students, she did not come from a particularly affluent background. Also, it would seem that there were family members who did not support her peripatetic academic lifestyle.

Vicinus notes that many educated women had more conflicts with their families than was common for the period. She adds: "Although few broke entirely with their families, professional women did not maintain as close ties with their relatives as those sisters who remained at home or married. The old family-based female network, primarily concerned with births, deaths, and marriages, was inevitably less important to ambitious single women seeking to make their mark in the larger world."[16]

Brooks's request that Rutherford intercede with Peterson on her behalf can partially be interpreted in terms of Peterson's well-known aloofness.[17] Rutherford must have sent her a very encouraging reply, since in her next letter Brooks discusses what she should do at Cambridge:

I am so glad that you are all so encouraging about the Fellowship, I have quite decided to take advantage of it at once. I am very grateful to you for your efforts on my behalf with Dr. Peterson. I shall want your advice with regard to the courses which it will be best for me to take but I can get that when I come home in June. Don't you think it would be a good thing for me to do some work at the Cavendish if they will let me? I think it would be much better for me than at the Newnham lab. If I enter a college at all it will be Newnham. I should like to go there if possible but I believe it is rather difficult to do so at short application. There is a Hall of residence for post-grad students to which I thought I might go but I do not imagine it will be nearly as pleasant as a college. Do you know of anything in the way of a scholarship that I could get there to increase the value of my fellowship? I am afraid that the fact of my holding that may debar me from competition for anything else. Of course I can get all I need from my family but I should very much prefer to be able to provide for myself. I know it will take nearer $1000 than $500 to put me thro'.[18]

The choice of the Cavendish Laboratory at Cambridge was most appropriate, as it housed the research group of J.J. Thomson, a collection of physicists working in the forefront of studies of radioactivity. Newnham College, the second of the women's constituent colleges at Cambridge University, had its own facilities but would not have had the specialized physics facilities of the Cavendish.

Brooks also remarks that she could get from the family as much money as she needed – a different attitude from her previous letter. Perhaps between the two letters to Rutherford she had received an encouraging response from her family. The source of the money could have been Edith's husband, Charles Gordon, who seemed to act as a family benefactor.

Principal Peterson of McGill sent Brooks his congratulations and told her that he looked forward to the probability of her applying for work at McGill after her year at Cambridge.[19] In the same month, the professor of mathematics at Bryn Mawr, James Harkness, wrote to Rutherford mentioning the award: "McKenzie and I were very pleased at Miss Brooks getting the Fellowship. I must say I envy her the year under J.J. Ions are in the air here: is it action at a distance from Montreal, or the more direct influence of Miss Brooks?"[20] Harkness was Brooks's instructor for the mathematics analysis course, while Arthur S. McKenzie taught Brooks in sound in relation to music, spectrum analysis, optics, and research laboratory work.[21]

Brooks wrote again to Rutherford, discussing her research results at Bryn Mawr and asking him to help her obtain permission to work in the Cavendish Laboratory with J.J. Thomson. She concludes with the remark "I shall try not to prove myself too dilettante if they allow me to do so."[22]

Thomson must have received a letter from Rutherford (though it has not survived),[23] for Thomson wrote to Rutherford: "I shall be very glad to give Miss Brooks permission to work in the Laboratory and to attend lectures and I am sure my wife and I will do all we can to make her stay in Cambridge pleasant and profitable. If she would like to live in Newnham College rather than in the town I will sound the authorities and see if it can be managed."[24]

It seems that Rutherford passed this information on to Brooks, for she wrote to him soon thereafter, expressing her gratitude to Rutherford and to Thomson. She adds: "It will be so cheering to be somewhere again where I think people know a great deal"[25] – a cutting commentary on the physicists at Bryn Mawr, though there was little high-quality physics research anywhere in the United States at the time.[26]

The situation regarding accommodation was less promising. In the same letter to Rutherford, Brooks remarks: "I wrote to Newnham some time ago and received a letter from Mrs. Sidgwick a few days ago.[27] She says I very possibly have to go into lodgings and just take my meals at the college as they have a great many applicants but they will make room for me if they can possibly."[28]

Brooks journeyed home to Montreal in June. From there, she took one of the frequent boats to England to commence her studies at Cambridge.

CHAPTER 5

LIFE WITH J.J. AT THE CAVENDISH

The Cavendish Laboratory was opened in 1874. The money behind the enterprise had been donated by the seventh duke of Devonshire, William Cavendish, for "a building for the teaching of Experimental Physics."[1] The construction of this building marked the change from the classical small laboratory of the professor and his student to the large research laboratories of the twentieth century.[2]

Although referred to as the Cavendish Laboratory, it was not just a single laboratory. Instead, it was a massive, three-storey building, containing research laboratories for different fields of physics research, a large general laboratory for students, a large lecture room for about 180 students, together with storage and support facilities.

While still at Bryn Mawr College, Brooks had expressed a desire to enter Newnham College. There were two women's constituent colleges at Cambridge: Girton and Newnham. Newnham was the newer one, having been founded in 1875,[3] and it had a stronger leaning towards the sciences, certainly a factor in Brooks's choice. The college was located in the town of Newnham, just across the River Cam from the rest of the university. It would have been a distance of about 400 metres from Newnham College to the Cavendish Laboratory, a convenient location for Brooks. Girton, on the other hand, was built a considerable distance outside Cambridge.[4]

Brooks was accepted as a member of Newnham College at a meeting of the General Committee on 1 May 1902. Although in her earlier letters Brooks had expressed a hope of living in the residences of Newnham, this must have proved impossible, as there is no record of this in the Newnham Archives.[5] According to her letters of this period, Brooks's mailing address was "Melrose" on Grange Road, a street that runs along one side of the Newnham campus.

As well as doing research, Brooks took several courses at the Cavendish.[6] In the Michaelmas (fall) term, she took Constitution of Matter and Properties of Matter, and in the Lent (winter) term, she took Discharge through Gases and Electricity and Magnetism. All of these courses were given by the illustrious J.J. Thomson himself. Brooks's work is noted as "research on the excited radiations from radium and thorium," and it involved about twenty-five hours per week in the Cavendish Laboratory.

We are fortunate to have several accounts of what it was like to work with J.J. Thomson. Professor Bumstead of Yale University, who worked with him at the Cavendish in the following year, expressed amazement that anything could be accomplished in a laboratory where no one arrived before 10 AM and all departed by 6 PM. He remarked that the laboratory was dominated by the personality of "J.J." Friendliness and helpfulness were common among the workers, but it was the practice in the laboratory to "jump into a fellow-student, if you thought him wrong, and to prove him wrong. In a good many places friendship does not stand that strain, but it usually does at the Cavendish."[7]

This aggressiveness could not have been pleasant for Brooks, considering her self-effacing attitude. Widnall notes that many women feel uncomfortable with the combative style of communication within research groups. She adds that some women suffer a permanent loss in self-esteem from vitriolic verbal exchanges.[8] This raises the possibility that it was a less-than-pleasant experience at the Cavendish that led Brooks to give up the idea of completing a doctoral degree.

Sir George Paget Thomson, the son of J.J. Thomson, describes how researchers each worked on their own problem and were expected to construct their own apparatus.[9] In the early 1890s, Thomson started the Cavendish Physical Society, a very informal gathering that met every two weeks during the term. At each meeting, one of the research students had to describe some research work, usually his or her own, and after the presentation, there was a discussion of the work. The concept of such a meeting had originated in Germany,[10] and this was the first of such seminar-type sessions in England.

Before the presentation, a formal tea was organized by Mrs Thomson and "one or two ladies connected with the laboratory."[11] Probably Brooks, as a woman scientist, was one of those expected to help serve the tea, a "tradition" that persists to the present in some British laboratories.[12] Mrs Thomson (Rose Paget), a former undergraduate research student of Thomson's, ceased her academic work upon her engagement to him.[13] Another of Thomson's colleagues, A.J. Strutt, the 4th Lord Rayleigh, adds: "The institution of tea was, however, a good one, because many of the workers in the laboratory had had little or no lunch, and required some sustenance if they were to take active part."[14]

Rayleigh also describes the more Spartan afternoon tea that was held in J.J.'s room every day:

> The tea hour itself was in many ways the best time in the laboratory day. The tea itself had no special quality; the biscuits were unattractive in the extreme, and very dull; the conversation sparkled and scintillated and as a social function tea was an outstanding success. There seemed to be no subject in which J.J. was not interested and well informed; current politics, current fiction, drama, university sport, all these came under review. The conversation was not usually about physics, at least not in the technical aspects, though it often turned on the personalities or idiosyncrasies of scientific men in other countries, who were known personally to some of those present and by reputation to all. J.J. had something to say on nearly any subject that might turn up. He was a good raconteur, but also a good listener, and knew how to draw out even shy members of the company.[15]

Brooks's time at Cambridge is first mentioned in a letter from Rutherford to Thomson.[16] Brooks herself continued to keep in contact with Rutherford, and it would appear that she relied on him considerably for advice. "I must have written to you last term in one of my deluded moments since I seem to have given you the impression that my experiment was going well then. [This earlier letter has not survived.] My results were sorely muddled for a long time but I begin to have placed most of my difficulties now though I have solved very few of them."

She then goes on to discuss the details of her experiments. One interesting point is that she refers to radioactivity decreasing to one-half of its value in about a minute. This is certainly the first measurement of the half-life of the thorium emanation (radon-220), and it compares very well with the currently accepted value of 55.3 seconds. She adds:

I have tried to revive it after it had decayed by bubbling throu' water but it does not recover. I thought at first that the emanation tho' dead might still be able to produce radio-activity but it doesn't. I am rather at a stand still until J.J. can give the matter his consideration. His modified air is behaving in a very unorthodox manner too and he is so worried about it that he doesn't pay much attention to us. He seemed quite interested however when I told him about my discoveries the other day and promised to think of it but he hasn't yet done so to any effect.[17]

Thomson was distracted by his own problems. At the time, he was under the impression that air could be made radioactive by being bubbled through water (to give "modified air").[18] This possibility appealed to Thomson, as it suggested a parallel between radioactivity and electricity – phenomena that could be imparted to matter without changing the fundamental nature of the substance. It was only in March and April of 1903 that Thomson was finally becoming convinced that the release of radioactivity did result in the formation of a different element.[19] Up until this point, he had considered the release of emanation from thorium was the result of some obscure chemical process, leaving the thorium unchanged. Brooks's observations were impossible to interpret in terms of Thomson's picture of the universe, so in desperation she appealed to Rutherford for help: "Will you let me know what you think of my experiment in its new light if it is not too much trouble? I know you have enough to think of at McGill without being bothered with mine."[20]

Brooks obviously found Thomson's assistance less than satisfactory. Thomson visited the laboratory only once a day for about an hour to check on the progress of each researcher. Rayleigh comments: "J.J. on his daily rounds sometimes made very impracticable suggestions, and this was possibly due to his being overworked ... When J.J. was posed with difficult questions, he would sometimes say that he would think it over – but this was comparatively seldom."[21]

Thomson was a more aloof character than Rutherford, and it would appear that Brooks found it hard to communicate with him. It is not that Thomson objected to women students, for comments expressed in his correspondence suggest a quite tolerant, if condescending, attitude to female physics students. In a letter written in 1886 to Mrs H.F. Reid, a long-time friend and confidante, Thomson remarks: "I think you would be amused if you were here now to see my lectures – in my elementary one I have got a front row entirely consisting of young women and they take notes in the most praiseworthy and painstaking fashion, but the most extraordinary thing is that I have got one in my advanced lecture."[22]

In a similar vein, he wrote to R. Threlfall, an assistant from 1885 to 1886, concerning the admission of women:

> We are just at the commencement of a great attack which is being made by the supporters of women to secure for them full admission to all the privileges of the university ...
>
> I do not think myself that it would do the university very much harm, but it seems to me that it would be bad for the women to be tied hand and foot to our system, for from what I see of them at the laboratory I am sure they require a rather different course from men: for example they always do very well in the first [part] of the tripos, but make a most awful hash of the second, in fact I think in nineteen cases out of twenty they had much better not attempt [it].[23]

Even though Brooks felt she was getting little recognition from Thomson, Thomson wrote to Rutherford, commenting: "Miss Brooks has been getting some interesting results with thorium, she finds that the emanation is able to produce induced radioactivity long after it has lost its power of ionising the gas around it."[24]

Rutherford must have replied to Brooks, for a subsequent undated letter from her acknowledges his help:

Thank you so much for having taken so much trouble about my experiment, it was very good indeed of you and I hope you did not think me too unreasonable to have asked you about it.

I am afraid I am a terrible bungler in research work, this is so extremely interesting and I am getting along so slowly and so blunderingly with it. I think I shall have to give it up after this year, there are so many other people who can do it so much better and in so much less time than I that I do not think my small efforts will ever be missed. I had an offer of the principalship of a girl's school in Halifax the other day and I have said I would accept if they will give me the salary I ask, I am rather doubtful if they will do it however. I have not heard anything definite about a post at McGill and I feel as if I would rather do anything than have to manoeuvre with Dr. Peterson about one, he is so very unsatisfactory. I had written to Miss Oakley before this other one was offered me asking about what was likely to turn up there but I had not heard anything from her.[25]

The address on this letter was Berlin, Germany, but there is no clue as to why she chose that particular city to visit. It could have simply been part of a European tour or perhaps a working visit, considering the leading role of the University of Berlin in physics research.[26] Brooks adds: "I left Cambridge two or three days ago and I expect to be on the continent for three or four weeks. From here I am going with a friend to spend Easter in the north of Italy."

The identity of the friend is a mystery, though it was possibly one of the two other women researchers at the Cavendish at that time: Jessie Slater[27] and Edith Wilcock.[28] Slater worked on topics closely related to Brooks's interests; hence, it is quite likely that in the male-dominated Cavendish Laboratory, they would develop a friendship.[29]

There is one particular item from Brooks's memorabilia that relates to her ties with the Cavendish. This is a humorous physics song written out in Rutherford's hand.[30] The song was one of several written for the annual December dinner of the research students from the Cavendish Laboratory.[31] Many were written by A.A. Robb, a mathematician from St John's College, Cambridge, and the song in Brooks's possession had his name appended.

The actual song was written somewhat later than Brooks's sojourn at the Cavendish, as it was first sung at the 1908 dinner in honour of Ernest Rutherford when he obtained the Nobel Prize for chemistry.[32] It is possible that Brooks travelled to England again for this most important occasion. Alternatively, Rutherford might have written to her about the event, but if this was the case, the letter has disappeared. The words of the version in Brooks's possession differ significantly from those published in the *Post-Prandial Proceedings*, but only slightly from a version Rutherford sent to Bertram Boltwood.[33] It would seem likely that Brooks's handwritten copy represents the original lyrics, while the published song was a revised and "improved" set of verses. Brooks's version is given in Appendix 2.

After her return from Europe, Brooks must have made plans to work again at McGill. As indirect evidence, the principal of McGill, Peterson, wrote to Rutherford and his colleague, H. Marshall Tory, lecturer in mathematics, asking what work could be assigned to her.[34] Her formal appointment as non-resident tutor in mathematics and physics at the Royal Victoria College was approved by the Board of Governors.[35] Peterson wrote to Brooks, telling her of the good news:

> I am glad to hear from you, and am happy to offer you a non-resident tutorship in connection with the Royal Victoria College, and other work in Mathematics and Physics, at a salary of $750 for the whole session – that is to say, including the months of May and June.[36] You are quite right in defining the work as something of the same nature as that which you were doing with us two years ago. The message that you have received through Professor Rutherford was sent to you as soon as I heard from the staff what kind of opening they could make for you.[37]

Thus, Brooks returned again to her favourite place – McGill University.

CHAPTER 6

BACK HOME TO McGILL

Once back at McGill, Brooks resumed her work as a tutor at the Royal Victoria College and rejoined Rutherford's research group. We have little direct knowledge of her life during this period, but we know that her research work relied heavily on measurements of the small electric charges induced by radioactivity. Such measurements were much easier to do in the damp climate of England than in the very dry winters of Montreal. Rutherford discussed the problem of electrostatic charge caused by walking across the varnished wood floors of the laboratory, and he devised a novel solution: "To reduce these disturbances I subsequently covered the floor with thin sheets of metal connected with earth, but, as we were using a high-potential battery with one pole earthed, some disconcerting experiences of acting as a conductor for a thousand-volt battery led me to abandon this form of screening."[1] This would certainly have been an unpleasant experience for the experimenter! Brooks herself remarked on the necessity of sitting absolutely still for extended lengths of time while making the measurements.[2]

Otto Hahn worked at McGill from 1905 to 1906,[3] and he described how each researcher constructed his or her own apparatus from simple components. The electroscopes were built out of large tin cans, on which were placed smaller tobacco or cigarette tins; the insulation of the electroscope leaf was made of sulfur, as they lacked the more conventional amber. This "string and sealing wax" attitude to the construction of equipment might have been a consequence of Rutherford's British training, where homemade apparatus was the trademark of the laboratory well into the twentieth century. It was also customary to work long hours, as Hahn adds: "Rutherford's enthusiasm and abounding vigour naturally affected us all. To work in the laboratory in the evening was the rule rather than the exception ... Frequently we would spend the evening in his house, where naturally little but 'shop' was talked, not always to the pleasure of the hospitable Mrs. Rutherford, who would have preferred to play the piano."[4]

On the basis of some of her work at McGill, Brooks published an article entitled "A Volatile Product from Radium."[5] This research was concerned with the polonium produced when radon, itself, decays. Frederick Soddy vividly describes this transformation in his subsequent textbook: "Every second, two out of every million of the atoms of emanation [radon] disintegrate, expelling α-particles and leaving a solid residue [polonium], so there is a sort of invisible snowstorm silently going on covering every available surface with this invisible, unweighable, but intensely radioactive deposit."[6]

Brooks deposited a thin layer of polonium (radium A) on the surface of a copper plate. She found that if the plate was placed in a testing vessel and then removed, the inside of the testing vessel became radioactive. She attributed this phenomenon to volatility of the decay product from the polonium. In other words, she concluded that some of the decay product had vaporized off the copper and then condensed on the walls of the vessel.

We now realize that even though Brooks's explanation was erroneous, her observation was very significant. According to the classical law of conservation of energy, when an α-particle is ejected from a radioactive atom, the atom should move in the opposite direction. Some of the atoms produced would be ejected from the surface of the copper and would settle on the vessel wall. A.S. Eve (Brooks's brother-in-law) compared the phenomenon to the recoil of a gun firing a shot.[7]

Rutherford recounted Brooks's findings at the Bakerian Lecture delivered in England on 19 May 1904.[8] He initially agreed with her explanation that the observed phenomenon was due to the volatility of radium B. However, shortly after Rutherford had presented the results, J. Larmor, secretary of the Royal Society, wrote to him, pointing out the "terrific kick back" that the atom would receive when an α-particle with one-tenth the velocity of light was ejected.[9] Rutherford changed his views and adopted this new explanation.

Brooks's observation of the recoil of the radioactive atom was largely overlooked by other scientists. When the research teams of Hahn/Meitner and Russ/Markower rediscovered the phenomenon in 1908, they were unaware of Brooks's earlier work.[10] As a result, Brooks received little credit for her original discovery.[11] These later physicists showed how the recoil effect could be used to separate off products from a radioactive decay. By transferring them to a new surface, one could study the products without further interference from the parent nuclei. A number of new radioactive elements and isotopes were discovered using the recoil method.

Otto Hahn announced his discovery of the recoil effect at a meeting of the German Physical Society in 1908.[12] Rutherford heard of this report and wrote to Hahn: "By the way, I thought I had the idea of the removal of atoms by recoil in my *Radioactivity* somewhere – see page 392 2nd edition. It is given in explanation of the volatility of Radium B observed by Miss Brooks. Send me a copy of your paper as soon as it is out."[13] Hahn noted that he was unfamiliar with Brooks's work but that he had consulted the report in Rutherford's text.[14] However, ever ready to claim credit for new discoveries, Hahn argued (unconvincingly) that Brooks's observations were not those of the recoil phenomenon.[15]

Another paper Brooks wrote during this period was "The Decay of the Excited Radioactivity from Thorium, Radium, and Actinium."[16] This comprehensive comparison of radioactive decay should be recognized as one of Brooks's most impressive contributions to physics. In the eleven-page paper, Brooks reports her study on the radioactivity of the decay product from radon. She shows that the rise and fall of the level of radiation can best be interpreted in terms of two successive radioactive changes.

Her work in 1901 had led to the conclusion that an element was transformed during a radioactive change. These later studies showed that the product was itself radioactive and, in turn, decayed. The discovery that there was more than one transmutation proved to be equally novel and important in untangling the complexities of radioactive decay. Rutherford's presentation at the Bakerian Lecture, "The Succession of Changes in Radioactive Bodies," leaned heavily on Brooks's research, making frequent references to the work of "Miss Brooks."[17] His report gained him great acclaim, but unfortunately Brooks's role seems to have become overlooked with time.[18]

We now realize the complexity of the radioactive transformations that Brooks was studying. For a start, different isotopes of radon are produced from the decay of thorium, radium, and actinium. These three isotopes of radon go through different series of transformations before they reach a stable isotope of lead. The changes are shown in Appendix 1. The work must have been started at the Cavendish, for Brooks acknowledges: "I am indebted to Prof. J.J. Thomson for his kind interest and assistance in a portion of the work done in the Cavendish Laboratory, and to Prof. Rutherford for his direction throughout the whole course of the investigation."[19]

Nevertheless, the fact that most of Brooks's publications gave her name as sole author would suggest that Rutherford regarded her as an independent co-worker. Her contributions were noted in an article written in 1906 by A.S. Eve on physics at McGill: "Reference may be made to some of the work done by research students. Miss Brooks has published several papers on various radio-active phenomena, and this lady was one of the most successful and industrious workers in the early days of the investigation of the subject."[20]

Brooks was elected to membership in the McGill Physical Society (based on the Cavendish model) on 8 October 1903.[21] The society had been established in 1897 with the objectives of discussing original work in progress at the Macdonald Physics Building, reporting on current scientific periodicals, and promoting physics research at the university. Any member of McGill University, interested in physical science was eligible for membership. It is curious that Brooks was not elected to membership during her earlier time at McGill. Robert K. McClung, her contemporary, had been elected a member of the society in October of 1899. It may have been that the membership of the 1898–1901 period were unwilling to accept a woman member or simply that Brooks never applied to be admitted, perhaps owing to a lack of confidence or an unwillingness to break into a male-only environment. It was only in 1902 that the first women, Miss Gates, Miss Dover, and Miss Marcuse, were elected to membership. Dover and Marcuse were both chemists,[22] while Gates was an American postgraduate researcher with Rutherford. Gates travelled even more than Brooks, though Gates's research on radioactivity proved to be of only minor significance.[23]

At the society's monthly meetings, it was customary to give presentations on one's research work or on important articles in the literature. This was followed by a discussion of the topic. At the 26 November 1903 meeting, the society minutes note: "The principal paper of the afternoon was an interesting account by Miss Brooks of an investigation by von Leich on the Chemical Action of Radioactive Matter in Solution."[24] Brooks was listed as being present at all the subsequent meetings of the society for that academic year.

Brooks applied for a position at Barnard College in the spring of 1904. There is no record of the reason for her leaving McGill, although it is feasible that she was seeking a more permanent position. As we shall see in the next chapter, there is also the possibility of a romantic attachment. Brooks obtained the appointment, and Principal Peterson wrote a pleasant response to her letter of resignation.[25]

This ended Brooks's link with McGill. It is likely that sometime in the summer of 1904 she moved to New York City to take up the appointment at Barnard College.

CHAPTER 7

THE RIGHTS OF A WOMAN: BARNARD COLLEGE

Barnard College was founded as a women's college in central New York City in 1889. It was instituted as an adjunct of Columbia University, with the dean of Barnard reporting to the president of Columbia. Barnard initially emphasized its connection with the cosmopolitan culture of New York, though over time it came to more closely resemble the other women's colleges. As Horowitz remarks: "By the turn of the century Barnard College had developed a unique blend of women's college and urban university. Barnard collegians enjoyed teas and dances with Columbia students, giving them a taste of coeducational university life. As seniors, they had classes with Columbia men."[1]

The physics department was on the second floor of Fiske Hall. The style of the building has been described as "Henry II, in overbaked brick, with limestone and terracotta trimmings."[2] The physics facilities consisted of a lecture room, a photographic dark room, laboratories for mechanics, heat, sound and light – one for the introductory work – and, in the basement, an electrical laboratory.

Brooks was appointed tutor in physics for 1904–1905 with a salary of $1,000.[3] In the fall of 1904, she taught mechanics and joint-taught the general physics course with Professor Margaret Maltby, the other member of the physics department. The following (winter) semester, she taught a course on light as well as the electricity and magnetism course. Each of the specialty courses had two hours of lectures per week and four hours of laboratory.

Brooks was reappointed for the 1905–1906 year at a salary of $1,100. Her teaching load for this second year was the same as in her first. No correspondence with Rutherford exists for this period, nor did she publish any work; however, there is an indication that she continued to do some research.

Her contract was renewed for the 1906–1907 year, but a complication occurred in the summer of 1906 – her planned marriage to a Mr Davis of Columbia University. Bergen Davis was a physicist who had obtained a PhD at Columbia in 1901, then spent a year at Göttingen, followed by a year at the Cavendish with J.J. Thomson. From 1903 to 1906, he was a tutor at Columbia University. Since Brooks and Davis had worked at the Cavendish with Thomson at the same time, it would seem certain that they had met there and likely that their relationship originated from this period. If true, this could account for Brooks's decision to leave McGill and take an appointment at Columbia's neighbour, Barnard College.

Davis was an eccentric character, lacking patience and sometimes tact. He had strong convictions and little tolerance for superficiality. Because of his dominant personality, he may have appealed to Brooks as the antithesis of her "weak and wayward" father. Certainly, such an attraction would parallel her lifelong friendship with Rutherford, another individual of strong views and brusque manner.

His colleagues considered Davis a widely read individual, his special interests being history, biography and poetry. "He especially enjoyed the contacts and discussions during the lunch hour at the Faculty Club. His menu never varied; he ordered it with a rubber stamp which read: 'Crackers and milk; Apple pie; Glass of milk; Bring it all at once.' Few lunches were complete without his bringing out a little notebook and a pencil stub to make some arithmetical computation suggested by the conversation."[4]

Brooks wrote to Dean Laura Gill to tell Gill about the impending marriage to Davis. The correspondence that ensued between Brooks and Gill gives us an insight into the problems of married women academics of the time.[5] The first letter from Brooks simply notified Dean Gill of her plans. It was written from the address of Brooks's parents in Montreal, 847 Hutchison Street. Brooks assumes that Gill has heard of the engagement from Marie Reimer, head of chemistry.[6] She tells the dean that her marriage, originally planned for June 1907, might now take place in September of 1906, though no reason for the change in date is given. Brooks notes that she has discussed her continued employment at Barnard with the head of physics, Margaret Maltby, who had had no objections, and she concludes: "Could you let me know if my being married will in any way invalidate my present agreement with the University or if you have any objections, as far as the college is concerned, to my retaining my position."[7]

This was not the first time that the Barnard administration had been faced with such a decision. The first dean of Barnard, Emily Smith, married George Putman in 1899.[8] After some discussion, the Board of Trustees allowed her to retain her position after marriage. This decision was quite exceptional for the time; resignation was usually expected immediately.[9] However, the board did insist on her resignation upon announcement of her pregnancy in 1900. The new dean, Laura Gill, was a serious, conservative individual, and it is unfortunate that Brooks had to deal with her rather than Smith.

Gill responded to Brooks's enquiry stating that in her view marriage would definitely spell the end of Brooks's career at Barnard.[10] Brooks continued the correspondence from a summer abode in the Adirondack Mountains of New York. Her next letter was a stirring response in which she expresses surprise that a resignation should be expected, though it is unclear whether her surprise was owing to her knowledge of the Smith precedent or to her own naïvety. Certainly it was common practice at the time to expect a woman to terminate her academic post upon marriage. Brooks adds:

I am quite sure that my duties will be, if anything, better performed under the new conditions but if I find that my new relationship, at any time, interferes in the slightest degree with my professional work, I shall, of course, at once tender my resignation. But failing such an outcome I should not be justified in resigning.

I think also it is a duty I owe to my profession and to my sex to show that a woman has a right to the practice of her profession and cannot be condemned to abandon it merely because she marries. I cannot conceive how women's colleges, inviting and encouraging women to enter professions can be justly founded or maintained denying such a principle.

I am sorry to have thus to appeal from your decision but I cannot acquiesce without violating my deepest convictions of my rights.[11]

In her reply, Gill invokes the views of the trustees: "I know that they expect and wish that a woman should dignify her home-making into a profession, and not assume that she can carry on two full professions at a time. This is more than most able men assume to be possible."[12]

The trustees may have changed their views after the Smith affair, or Gill may have been selective in those she consulted. Virginia Gildersleeve, who followed Gill as dean, commented that Gill was an able administrator but was somewhat tactless.[13] Although the dean was officially responsible to the president, Dr Nicholas Butler, Gill tended to enlist the sympathy and support of the trustees in matters where her views were at odds with Butler's. During the entire exchange of letters, Butler was away from campus, and although Gill had written to him about the issue, she does not mention any response.

In this same letter, the dean mentions the "detrimental publicity and unpleasantness" that a protest by Brooks may cause the college and, by inference, the harm that it might do Brooks's future career. Gill's subsequent remarks make her position clear: "I honestly believe that the good of the College and the dignity of the woman's place in the home demand that your marriage shall be a resignation."

At this point, the head of physics, Margaret Maltby, intervened. It is curious that she does not refer to the Smith precedent, as one assumes the case was known to both Maltby and Gill. Maltby pleads that Brooks's situation be regarded as a special case so that Barnard does not lose her exceptional talents:

She is greatly interested in her research and teaching. She enjoys both and she is a very good teacher. I know of no one who could take her place. She is efficient, helpful, loyal and interested in the work. I never worked with one I liked so well. She knows about the apparatus now, and we have our plans laid for development of the department for the next four years. So long as she and Mr Davis agree to and wish the continuation of her teaching I do not think one can do better than to retain her. She is well prepared for the work and her heart is in her work. I am confident she will not keep house at present, so house-work will not interfere.

It is not for the money she can earn – though that is a consideration, of course – but it is really because she wishes a means of expression, like a man, and the work is near to her heart. I think I can sympathize with her thoroughly. Neither you nor I would like to give up our active professional lives suddenly for domestic life or even for research alone. She has bent every energy towards the end of preparing herself to teach and to be a physicist. I think if she wishes to continue for two or three years her teaching and can devote her energy to it as heretofore, we would make a mistake to refuse her the opportunity.

Maltby concludes with a reiteration of Brooks's talents:

I know of no woman to take her place – no one available who has the preparation and the personality and ability to teach, and the skill in physical manipulation, that she has. She has the English view of the matter. You know that many English women after marriage have done valuable scientific work or took in other fields. I believe Miss Brooks's life will be richer and happier for her continuance in the teaching for a few years at least.[14]

Maltby, in tendering this advice, spoke with considerable moral and professional authority. An outstanding scientist, she had obtained a BA and MA from Oberlin College (in 1882 and 1891, respectively) and a BS from the Massachusetts Institute of Technology (1891). She was the first American woman to obtain a PhD at Göttingen University, which she accomplished in 1895 working with the famous physical chemist, Kohlrausch. Maltby later did research in theoretical physics before committing herself to teaching at Barnard College (1900–31).[15]

Gill's response to Maltby (30 July 1906) is unyielding, quoting the comments of one of the trustees: "The College cannot afford to have women on the staff to whom the college work is secondary; the college is not willing to stamp with approval a woman to whom self-elected home duties can be secondary." Gill continues the letter with an expansion of her own views: "There are many phases of scientific, educational, and philanthropic work which a married woman can do without possible embarrassment or business disadvantage. Regular class-room work which cannot be interrupted without serious detriment to the class is not a form which is, in my judgment, adapted to young married women."[16]

For a contrasting view, it is interesting to quote from the experiences of Gill's successor, Gildersleeve:

Soon after I became Dean [1910] a young woman who was an assistant in Physics came to my office to tell me she expected to be married the following summer. "How nice," said I, "I wish you much happiness." "But," said she with a troubled look, "I should like to go on here next year as assistant in Physics, as I've already been appointed." "Well, why not?" I asked in some perplexity. "There's no objection from the College?" she queried. "None at all," I answered, "so far as I can see. As long as you do your work here satisfactorily, your marriage seems to me entirely your personal business."[17]

It is astonishing how this case paralleled that of Brooks. The assistant in physics to whom Gildersleeve referred was Grace Langford.[18] Langford was well aware of the experience of her friend and predecessor, Harriet Brooks. Gildersleeve then reviews the Dean Smith case (but does not mention the Brooks precedent) and adds:

Ignorant of all this, I had simply not thought of asking the Trustees about the case of the young assistant in Physics, but had embarked on this policy quite on my own responsibility. We continued to have married women on our staff fairly often; no Trustee objected. From my observations, I began to realize more vividly the difficulty of combining marriage, children, and a professional career; but I realized also that in some cases the combination absolutely had to be made.

Brooks's engagement to Davis was terminated about the beginning of August 1906, though no definitive reasons have come to light. In a short note, she wrote to inform Gill that her marriage had been postponed indefinitely and to request continuance of her position.[19] However, within a month Brooks submitted her resignation, the tone of which would indicate that she was suffering from considerable nervous strain as a consequence of the broken engagement.

> I find that it will be almost impossible for me to fulfil my engagement at Barnard for the coming year owing to the unfortunate necessity I have been under, of terminating my engagement to be married. If it is possible I should like to have my place filled and I shall spend the year in study abroad. I am very sorry indeed to have to ask this favour at so late a date but I think that it is better than to undertake a task under conditions which render its successful accomplishment so doubtful.

Perhaps with the close ties between Barnard and Columbia, Brooks would have had to continue a working relationship with Davis, for she concludes with the comment: "If you think it quite impractical or if no one can be found to undertake the work, I shall of course go on with it but it will be a very difficult and painful year for me."[20]

Brooks may have found Davis's dominant personality unbearable, for in the postscript of a letter to Hahn, Rutherford remarks: "You may be interested to hear that Miss Brooks has broken off her engagement, got tired of the young man!!"[21]

Maltby wrote Gill with some additional comments about the broken engagement: "Late last evening I received a letter from Miss Brooks in which she states that she feels she cannot come back to Barnard next year. She wishes to go abroad. You know that the engagement with Mr. Davis is broken. It is very unfortunate for both that there was ever an engagement. She feels that the nervous strain of coming back at present is too great."[22]

The formal acceptance of Brooks's resignation came from Gill's secretary. This may indicate Gill was still angry with Brooks over her original engagement or because of her sudden resignation just before the start of the academic year. The letter states that in view of the circumstances, Gill will not hold Brooks to her contract and will regretfully accept her resignation.[23]

Brooks's final letter in this exchange comes once more from the Montreal address, thanking Gill for her kindness and adding in parting: "I probably expressed myself with undue acerbity in the case you mentioned but I still strongly feel that it is not quite a fair attitude on the part of the college, that altho' it might, with justice, be the general policy, particular circumstances might cause its enforcement to be a great injustice."[24]

From the academic world of Barnard, Brooks's life took a new turn, bringing her into contact with a completely new circle of acquaintances.

CHAPTER 8

A SUMMER IN THE ADIRONDACKS

Three of the five letters that Brooks sent to Gill had the address of Summerbrook, Hurricane Ridge, New York. Her residence at this house in the Adirondacks marked a new direction in her life. Summerbrook was established by Prestonia Mann in 1896 to be operated by groups of social workers, settlement people, and writers on reform subjects in imitation of the Brooks Farm[1] (this Brooks was no relation to Harriet). The original Brooks Farm was a Utopian collectivist community founded by George Ripley. It was operated by Nathaniel Hawthorne, Ralph Waldo Emerson, Margaret Fuller, and other Boston intellectuals in the 1840s[2] according to the teachings of the French Utopian Charles Fourier.[3]

Prestonia Mann Martin came to play an influential role in Brooks's life. In fact, Brooks's letter to Gill on her rights to pursue her career resembles the passionate style of Martin more closely than it does the more circumspect style of Brooks herself.

Martin came from an affluent Staten Island family who were themselves social reformers. She was described as being rich, cultured, leftish, brilliant, altruistic, frank, lovable, grand, and eccentric.[4] The Mann family had attended the lectures of the philosopher Thomas Davidson at Glenmore, East Hill, in the Adirondacks, and it was here that Prestonia obtained the idea for starting her own school. She purchased land from the prominent socialist and then-professor at Chicago University John Dewey less than half a mile from Glenmore, on the same Hurricane Ridge. As Porter remarks:

She determined to foster the more humane aspects of society and to make her school a pattern of the enlarged family – the ideal community. A large woman, friendly, confident, and determined, she gracefully introduced new customs in household management and dress and played Lady Bountiful for miles around. She led her guests in the consideration of social problems and in a two-hour labor stint. Nothing but physical labor counted as work with Mrs. Martin, but it was all made a cooperative game in which social reformers of New York and Boston took part – such people as Edwin Markham and his wife, Lillian Wall, Jane Addams, Henry D. Lloyd, Upton Sinclair, and Charlotte Perkins Gilman. Wash day was an occasion for hilarity, with the joining of hands and singing, "Glory, glory, and the wash goes marching on!" Flowing summer gowns, brightly decorated blouses, unusual male and female outfits went thru the wringer and popped buttons were lined up for recovery.

At Summerbrook, there was the two hours devoted to labor – cutting wood, tending the grounds, laundry. Each guest signed a labor book at night, telling what he or she had accomplished. There was time for study, for discussions which went on all day and far into the night, for daily reading of favorite passages by Prestonia and for recreation – climbs on Hurricane Mountain, drives in the pony cart, and long walks. There were no formal lectures at Summerbrook.[5]

John Martin, an English Fabian socialist,[6] came to lecture at Columbia University in 1899, and he joined the community at Summerbrook the same year. In the following year he married Prestonia Mann.[7] Prestonia Mann Martin wrote three books: *Prohibiting Poverty,*[8] an extremely popular book in its time, running to thirteen editions; *Is Mankind Advancing?*;[9] and a book of poems, *Riding Lessons on Pegasus.*[10] She also co-authored *Feminism, Its Fallacies and Follies* with John Martin.[11] Prestonia looked back to the Golden Age of Greece and expressed doubts about the modern times. In particular, she espoused "humanism" and abhorred feminism, which she perceived to threaten the foundations of society. She argued that it was the duty of intellectual women to marry and produce children, else the country would be swamped by the offspring of the masses. She commented on the poor health that she observed among college girls, a clear sign that the delicate feminine nature was unsuited to severe mental toil.

Prestonia demanded that future generations of children be raised exclusively in healthy farm environments and be required, at the conclusion of schooling, to perform several years of compulsory state service. Her arguments exhibit a mélange of socialism, anti-feminism, racism, and fascism. Prestonia believed that to lead a productive life, a woman should do three things: write a book, have three children, and plant ten thousand trees.[12] She lived up to her ideals, with three and a half books published, three (adopted) children, and enough trees planted at Summerbrook to turn open country into a forest.

The occasion when Prestonia and Harriet Brooks met at Summerbrook was alluded to in a presentation given during the 1950s by John Martin: "a very good friend of Prestonia's whom she acquired through her experiences with the tired teacher's [camp]; a lady who was an eminent physicist."[13] It seems clear that Martin was referring to Brooks. The stress of Brooks's contretemps with Gill might well have caused her to seek the peace of the Adirondacks. However, in later life, John Martin's recollections were sometimes faulty, and there is evidence that Brooks had met Prestonia Martin earlier that year in New York.

This earlier meeting involved Maria Andreyeva, the common-law second wife of Maxim Gorky (the pen name of Alexei Maximovich Peshkov).[14] Born in 1868, Gorky attained fame as the first literary chronicler of Russia's urban working class. He wrote several articles for the American Press, and after the tsarist regime imprisoned him in 1905 for his inflammatory writings, it was the outcry of American literary writers and journalists that influenced his release.[15] Gorky was encouraged to visit the United States to raise funds for the revolutionary cause, and he arrived there on 10 April 1906, together with Maria Federovna Andreyeva and Nikolai Burenin, his secretary. Burenin was a member of the Bolsheviks' Central Committee,[16] and it was said that his role was to prevent Gorky from deviating too far from accepted revolutionary policy.

The Gorky party mingled with the best of New York society, for in 1906 the word "revolution" was still acceptable to the average American.[17] The change in attitude to the Gorkys came about through two incidents that happened on 14 April. First, Gaylord Wiltshire sent a letter in Gorky's name expressing support for two imprisoned leaders of striking miners who had been accused as accessories in a bombing murder. This apparent interference in domestic politics was not well received. On the same day, the news broke that Gorky was not married to Maria Andreyeva and that his "real" wife was back in Russia.

In spite of the explanation that divorces were not permitted in Russia and a supporting telegram from his first wife, Ekaterina Pavlovna, making it clear that they had indeed separated permanently, Gorky's supporters vanished.[18] Hotel after hotel refused to accept him. John and Prestonia Martin were the only people to come to the rescue of the Gorky party, offering them the use of the second floor of their house at Grymes Hill, Staten Island.[19] At this time, Gorky was accompanied by Andreyeva, Burenin, and Zinovy Peshkov, Gorky's adopted son who had been living in New York for some time.[20]

While the Gorkys were residing with the Martins, John Dewey, professor of philosophy at Columbia University, hosted a meeting for Andreyeva. It was at this meeting that Brooks, still at Barnard College, could have first met Andreyeva and Prestonia Martin, for Dewey commented: "A group of women had asked my wife if she would give the use of our apartment to Mme. Andreyeva to speak on the condition of women in Russia; a sensational newspaper reported that this was a reception to her and the Barnard girls were the guests."[21]

The occasion was reported in more detail by the New York newspaper, *The World*: "All of Barnard College, from the faculty down to the fluffiest freshman, was wrought up yesterday over the reception by Prof. John Dewey, of Columbia, to a score or more Barnard girls where the guest of honor was Mme. Andreieva, the pretty, blond actress who travels with Maxim Gorky." The long article describes the outcry and impending investigation into how such an event was permitted. The feature concluded: "The reception that has made such a stir was carefully planned. Mme. Andreieva was escorted to the house by Mrs. John Martin of Rosebank, S.I., in whose home she and Gorky have been guests since the hotels of New York were closed against them. The Russian actress and revolutionist made an address in

French which was translated to the audience by Mrs. Martin. She was heartily applauded and seemed to enjoy greatly the levee which followed her speech."[22] The version reported by the *New York American and Journal* was much less inflammatory: "To meet the actress, there were asked about twenty-five of the college girls who were interested in the cause of Russian freedom."[23]

According to the memoirs of Andreyeva, Harriet Brooks as well as Ernest Rutherford met Gorky and Andreyeva at the home of the Martins on Staten Island.[24] Unfortunately, as no mention is made of why the meeting occurred, we cannot be sure that it was a result of the Barnard presentation. Andreyeva also made a year-by-year note of the important people that she met. For 1906, she writes, "America. Wells [H.G.]. Rutherford. Harriet Brooks. Martin. Dewey. Burenin and others."[25]

Gorky and Andreyeva moved with the Martins to Summerbrook in late June.[26] Brooks must have arrived there about two weeks later, for her letter to Gill of 10 July was sent from Montreal, while that of 18 July was sent from Summerbrook. It was quite easy to reach Hurricane Ridge from Montreal, as the New York and Hudson River Rail Road ran two trains each day from Montreal to New York via the Adirondacks.[27]

The Burenin's memoirs devote considerable space to the period at Summerbrook. He remarks:

> Then Prestonia Martin invited us to spend the summer with them on their estate located in the Adirondack Mountains, on the border with Canada. To go to this estate from the nearby town of Elizabethtown by horse was about 25 versts [16 ½ miles]. The Martin estate was composed of two large sections. One was in a depression, surrounded on all sides by mountains, and the other was on the slopes of Hurricane Mountain. The first section carried the name "Summer Brook," and the second "Arisponet."

Burenin described life at Summerbrook vividly:

> At Summer Brook we had lodged in a separate house. Each evening we gathered in the sitting-room, which took up more than half the house, with a large fireplace into which we threw half-burned logs. Through a large picture window on the opposite side of the room we could see the

night-time sky with bright stars and the dark silhouettes of the mountains which disclosed the horizon.

The large concert piano seemed small in this huge room, and we had to play by the light of dozens of candles put in iron candle holders that exceeded the height of a person. Somewhere in a dark corner gleamed the horrifying eyes of a great horned buffalo head, which hung on one of the walls.

In the room during those evenings, besides Maria Federovna and Zavolzhsky, sat our hosts: Prestonia Martin and John Martin. They played chess in deep thought, very amusingly calling each other by Russian names: "Prestonia Ivanova" and "Ivan Ivanovich." Sometimes Prestonia Ivanova would utter a shriek: "Damn you!" Ivan Ivanovich would answer this in Russian with a sharp English accent, "Spacibo, do sfidanya" [Thank you, good-bye], and remove one of her chess pieces.

Besides us, there was a professor from Columbia University in New York who stayed with the Martins, a Miss Harriet Brooks, who had already become famous for her important discoveries in the field of radio[activity].

None of them understood Russian, but they sat quietly, as if spellbound. It seemed that they perceived the stories of Gorky like music – not understanding the words, but guessing the meaning.

When Gorky fell silent, Prestonia Ivanova would exclaim with fervour: "I didn't understand a single word but it was splendid!" And everyone asked Andreyeva to translate what Gorky had discussed. Maria Fedorovna translated with great proficiency, fascinating her listeners, since she passed almost word for word what Gorky had said.

We sat up very late at night, the candles in their holders burned out, and the fireplace blazed up when the logs were stirred, going out and burning down, illuminating the room with an odd red glow. The people blended into the darkness and the voice of Gorky spoke, like a legend, about his past experiences, first painting the nature and beautiful spirit of the people, then freezing our souls with the heavy and cheerless things he had to live through in the adventure of his life.[28]

The Gorky Museum in Moscow has a number of photographs taken by Burenin in late August showing Gorky, Andreyeva, Peshkov, John and Prestonia Martin, and Brooks.[29] The background views include the sitting-room at Summerbrook, the porch of the house, the neighbourhood of Ausable Lake, and

Hurricane Mountain (the people and locations are identified on the back of the photographs). The current owners of the house possess a slide made from a photograph of Prestonia and Harriet Brooks sitting beside the big fireplace in the main house.[30]

Burenin recalled a particular incident that reflected the carefree life on the estate:

One Sunday, a day when the owners attended to themselves and the servants took a walk, we sat with all the company in our favourite sitting-room. The mail brought some good news from L.B. Krasnin about winning a large party financial deal. Sensitive Prestonia Martin noticed that something wonderful had happened and she began to congratulate her "Russian children," as she called us, and was ready to dance for joy for us. I sat at the grand piano and started to play a gay cakewalk. Zavolzhsky, very talentedly, began to imitate a Negro dance, and was joined by Harriet Brooks, and having started a dance in the house, they ran out in general laughter to the garden flooded by brilliant sunshine. In the garden a young fellow joined him – the gardener, but he was also the driver and a "jack-of-all-trades." Dancing around the house, they caught the housemaid, a Negress-cook, who picked up and carried Prestonia Martin, and they caught up with Gorky himself and Maria Fedorovna, who then carried Ivan Ivanovich, and it was thus that they started dancing some unfamiliar round dance – we, in the Russian manner, and the Americans, in their own.[31]

During this eventful summer, Brooks must have told Gorky of her problems at Barnard, for in his diary, Gorky had written part of a conversation:

Miss Brooks, a physics professor at Columbia College, is marrying Professor Davis of the same university. The head of the college, an old maid, announced that after the marriage Brooks dare not appear at the college.

Brooks My fellow professors are not protesting.
Gorky Why don't you protest?
Brooks It's impossible. They won't accept me anywhere. They say, this is the Brooks they wrote about in the papers.[32]

Clearly, Brooks believed (probably correctly) that if she had contested the termination of her contract, the publicity would have made any other college reluctant to hire her.

The guests stayed at Summerbrook until late in August, when members of the Fabian Society arrived to take up residence.[33] The Gorky party moved about a mile down the road to "Arisponet," a more secluded annex camp, the name being an anagram of Prestonia. The group was visited by the writer John Spargo, who commented on Arisponet: "The large living-room of the old Community, barren of ornament, the rough bare pine boards never having been covered served as living-room and study. The side benches were literally covered with new Russian books and magazines, mostly, I learned, of a scientific and philosophical nature. An upright piano in one corner of the room was the only sign of pleasure or luxury to be seen."[34]

Gorky wrote to his first wife, Ekaterina Pavlovna, in early September:

> I'm staying in a very deserted spot in a forest, 18 miles from Elizabethtown, the nearest town; but the Americans come out here to take a look at me. They are afraid to enter the house; to know me means to compromise themselves. They stroll about in the forest in anticipation of a casual encounter. There are five of us here; myself, Zina,[35] a Russian who arrived here with me as my secretary, a physics professor, Miss Brooks, and a nice old lady.[36] There are no servants, we do the cooking and everything else ourselves. I wash the dishes, Zina goes on horseback for provisions, the professor brews the tea, coffee, and so on.[37]

The fifth person was almost certainly Miss Jones, a teacher from Chicago, whom Burenin mentions as having been there at the time.

From the addresses on the later letters from Brooks to Gill, we know Brooks was at Summerbrook on 1 September but had returned to her address in Montreal by 13 September. It is possible that she travelled back to Montreal with Gorky and accompanied him to McGill University. According to information gleaned from the Gorky Museum, Gorky visited Rutherford's laboratory at

McGill in late September, probably in the company of Brooks.[38] Danin, in his biography of Rutherford, notes: "But how far outside this narrow circle he [Rutherford] felt when a visitor from Russia unexpectedly called upon his physics building. The visitor was not a physicist, but a writer. Having travelled all across Europe and America in 1906, Maxim Gorky arrived to meet him."[39] We have independent evidence that Rutherford and Gorky did meet at some point, for in a letter to B. Perott in 1935, Rutherford notes that he knew Gorky personally.[40]

Brooks returned to New York in mid-September and rejoined the Gorky party in preparation to travel with them to Europe. The *New York Times* of 13 October 1906 notes:

> Maxim Gorky, the Russian novelist and Socialist, sails today for Naples on the North German Lloyd steamer Prinzess Irene. Mme. Andrieva, the Russian actress, who accompanied him to this country, and four friends will go with him.
>
> Gorky gave a farewell dinner to some of his intimate Russian and American friends in the Louis XVI room of the Café Martin last night. Mme. Andrieva sat at the head of the table. Among the twelve guests were two women ... For the present their plans were to stay in Italy for several months and then go to Paris and perhaps later to London. After dinner the party drove to the Victoria Theatre, where two boxes had been engaged.[41]

The two women mentioned by the newspaper were most likely Prestonia Martin and Harriet Brooks. The article states that four others were travelling with the Gorkys, but according to Burenin's account of the voyage, Brooks and Burenin were the only ones to accompany them.

Another New York newspaper, *The Sun*, had a more abrupt announcement: "Maxim Gorky, the Russian revolutionist and writer, accompanied by his companion, Mme Andreieva, sailed from Hoboken yesterday on the steamship Prinzess Irene for Naples. Gorky's seventeen-year-old adopted son was among the party of a score that were at the pier to say goodbye to the pair."[42] One would assume that the Martins were on the dock as well, bidding farewell to their summer guests: Gorky, Andreyeva,

Burenin, and Brooks. Thus, at 11.00 AM on 14 October 1906, Brooks set off across the Atlantic to embark upon yet further new exploits.

CHAPTER 9

EUROPEAN TRAVELS: CAPRI, PARIS, AND LONDON

According to Burenin, a bad storm arose during the crossing, but even so, he took a moment amid the swells to remark on the sumptuous surroundings.

The large salon of the ship was covered with white furniture. Everywhere palm trees or lively flowers in multi-coloured majolican vases stood in their beauty, firmly rivetted to the floor. Original paintings by famous artists hung on the walls. Large mirrors with their own reflections greatly increased the size of the room. The bright sun flooded the entire salon and cheerfully reflected off glittering objects. There was a strong pitching storm, and huge masses of white foam and spray flooded the ship. In order to reach the inlaid concert piano, which was fastened to the floor, I had to search for support upon any object along the way. The majority of the passengers were ill with sea sickness. I had not escaped this fate. But Gorky and the violinist were unaffected. We started to play. The long bench I was sitting on alternately scooted toward the piano, and then away.[1]

There is a photograph in the Gorky Museum of Gorky, Andreyeva, and Brooks on the deck of the ship. Before giving his photographs to the museum, Burenin had added a commentary to each. On this particular one, he had noted: "The steamer completed its trip between New York and Naples, passed the Azores, through the Straits of Gibraltar."[2]

The arrival of the four travellers in Naples on 26 October 1906 is described by Bykovtseva as an event that caused considerable excitement.[3] The Italian newspaper *Avanti* carried a large number of articles concerning Gorky. The first one referred to the arrival of Gorky, his companion Andreyeva, his secretary, and a female friend on the *Prinzess Irene*.[4] The paper stated that they were occupying a suite at the Hôtel Vésuve. In the Gorky Museum are two photos taken by Burenin that show Gorky, Andreyeva, and Brooks first coming out of the hotel and then sitting in an open car.

Every activity of Gorky and his companions was reported in the Italian newspapers. The group clearly caused an uproar wherever it went. For example, on 27 October the four of them attended the Politeama theatre to see the play *Mascotte*.[5] Burenin comments: "At Gorky's request I took a lodge seat at the Politeama theatre in Naples. We were a little late and entered the theatre only when the overture had already begun ... Suddenly the lights went up in the hall, and the entire audience rose from their seats to the cry: 'Long live Gorky!' 'Long live the Russian Revolution!' 'Down with the Czar!'"[6]

Burenin also describes an incident that could have proved fatal to the entire party. On 28 October, Gorky was invited to an open-air meeting to be held the following day. The party of four journeyed from their hotel in a horse-drawn open landau, followed by hordes of Gorky's supporters and with a large number of police and guards on hand. About five thousand people attended the meeting, which was stirred by the revolutionary addresses of Gorky and the Italian socialists. Burenin describes the events that followed the meeting:

Aleksei Maximovich Gorky, surrounded by a large crowd singing workers' hymns, departed together with M.F. Andreyeva to Saint Gaetan Square. The immense throng of people gave him a thundering ovation. Miss Brooks, who had come with us from America, and I stuck close to them, and with difficulty we sat as a foursome in the waiting carriage.

Thousands of people accompanied us, but suddenly the road appeared blocked by a platoon of soldiers and a large number of *carabinieres*. The soldiers were given the order to fix their bayonets. This irritated the crowd and they closed in on the soldiers. Gorky rose in the carriage and looked with bewilderment on the two cordons of soldiers who were ready to fire on the crowd. The shouting grew in strength. The furious commissioner ordered the signal to prepare to fire. Maria Federovna, pale, stood before Gorky, and I tried to shield them both.[7]

According to Burenin, an elderly lawyer rushed up to the officer at the last minute and persuaded him not to give the order to fire. It must have been a heart-stopping few minutes for all in the carriage, particularly for Brooks, considering that Burenin seemed more intent on protecting the Gorkys than her.

Burenin was sent to investigate the island of Capri, as it was felt that the climate on the island would be better for Gorky's health,[8] though *Avanti* suggests it was for reasons of tranquillity.[9] The party moved to Capri, staying first at a hotel, but Maria Andreyeva insisted that a villa be rented to make Gorky's life "as comfortable and cozy as possible."[10] They moved to the Villa Blesus, which Troyat describes as follows:

Situated in the lower part of the isle of Capri, the house where Gorky stayed was large, opulently furnished and surrounded by a flower garden, with a magnificent view over the sea ...

Now and then he left his island and went to Naples, Florence, Rome, or Genoa, but he always came back to his home port. Increasingly, numerous visitors came to Capri to pay their respects to the master in his "gilded cage"; some were writers or artists, some were simply sightseers, and most were Marxists.[11]

Brooks kept four photographs in her memorabilia from this period. Two, taken by the sea, were of Brooks with Burenin and three young boys (Burenin, in his memoirs, notes that he befriended some little Italian boys who taught him how to swim). The third photo shows Brooks sitting at a table with Gorky, Andreyeva, Burenin, an elderly lady, and one other, who may be Leonid Andreev or Ivan P. Ladyzhnikov.[12] Andreev was a famous Russian writer, while Ladyzhnikov managed a book-publishing company in Berlin on behalf of the Bolshevik Central Committee. The final photo is of Gorky, wrapped in a cloak, melodramatically holding an old pistol. Burenin also took a number of photographs (now in the Gorky Museum) of Brooks, Gorky, Andreyeva, Burenin, Ladyzhnikov, and Leonid Andreev, on the terrace of the

Villa Blesus, in a small restaurant, and on a walk through Piccola Marina and Marina Grande.

Brooks also saved two newspaper clippings from this period. One item was an article by John Martin entitled "The Truth about Maxim Gorky" that appeared in the *New York Times* of Sunday, 24 February 1907. In this long article, Martin defends Gorky and, in particular, his relationship with Andreyeva. He complains bitterly about the treatment accorded the Gorkys by the pens of New York journalists. The second clipping came from the Paris Communist newspaper *L'Humanité*; it consisted of two letters from Gorky under the heading "To My Detractors!" Both were dated 29 October 1907 from Capri. Their contents were a rebuttal of the French press's virulent attacks against him.

It is part of the Brooks family history that Burenin had proposed marriage to her,[13] but there is no substantiation of this in Burenin's memoirs. However, in the Gorky Museum there is a portrait of Brooks with the inscription (in Russian) "Henrieta Georgievna Sept 3 1906. To Burenin." In the conventional style of Russian names, Brooks would have adopted the name of her father (George) and added the feminine ending (-evna), hence Georgievna.

Despite the lure of the lifestyle on Capri, Brooks must have felt the need to resume work. She travelled to Paris and spent part of the 1906–1907 year working with Marie Curie. Unfortunately, no documentation survives to indicate how she obtained a position at the Curie laboratory. This could have been arranged by Rutherford or even by Gorky. Gorky was quite fascinated by the phenomenon of radioactivity and had visited the Curie laboratory himself in the spring of 1906. In Gorky's memoirs, Fedin writes: "He was telling me with genuine gratitude about the Curies whom he had met personally and whose laboratory he had visited when he lived in France."[14]

Brooks must have returned to Capri for the New Year celebrations, for in his memoirs Burenin remarks that he organized a New Year's concert by the local boys that bought Andreyeva, Brooks, and Gorky out of the house to discover the source of the noise. A letter from Andreyeva to I.P. Ladyzhnikov of 4 February 1907 indicates that Brooks was again in Italy at the beginning of that month: "And also Harriet, she is here with me in Naples, in a few days she is going to Paris to visit Curie."[15] Andreyeva also passed along Brooks's regards.

There is disappointingly little about Brooks in the Curie Institute archives, only an entry in the 1906–1907 ledger of the Laboratoire Curie[16] to the effect that Miss Brooks was an independent worker (*travailleur libre*) with André Debierne.[17] The dates and duration of her stay were not noted.

Pierre Curie had been killed in an accident on 19 April 1906, and it was only in the fall of 1906 that Marie Sklodowska Curie was given a formal appointment. Thus, she was just starting to direct research work on her own at this time. In this, Marie Curie was helped greatly by the continued presence of André Debierne, a co-worker and close friend.[18]

Surprisingly, there seem to be only two accounts of life at the Curie Institute. This description is given by Ellen Gleditsch, who joined Curie's research group in 1907:

> There was not much room, we were only five or six workers. Marie Curie came every day and spent many hours there ...
>
> She had a thorough knowledge of the work of each student. When she visited the research rooms, she stopped with one researcher to ask about the details of their research; with another to give a word of encouragement; here to re-do a calculation; there to criticize an apparatus; farther on to admire another piece of equipment; always up-to-date with what one was doing, always full of interest for all the details ...
>
> In the laboratory, her face, which was usually serious and slightly sad, became animated, she smiled often, and one even heard a fresh, young laugh.[19]

This image of Curie's personality (possibly coloured by Gleditsch's friendship) is in sharp contrast to the assembled information of Badash, who notes that the saint-like image of "our lady of radium" owes more to the media than to her real personality; according to his evidence, she was "stubborn, oversensitive, cold, conveniently ill."[20]

In 1908, May Sybil Leslie joined Curie's group. Her comments highlight the nonchalance with which radioactivity was treated at the time, and they also describe the key role that André Debierne played in the running of the research group.

The laboratory consists of a collection of small rooms. The Salle à Chemie in which Mme. Curie does most of her chemical work is detached from the main building & has by this time become very radioactive.

Mme. Curie spends a considerable amount of time working in the laboratory. She does not appear to come around much to the students but receives them very kindly when they seek her. She does not speak English at all, nor does she appear to understand spoken English except a few scientific terms. She speaks very quickly and to the point & is very quiet in manner but by no means languid. She has a face of great intelligence and the expression is rather warm and sad in general, but she has a most charming smile at times which quite transfigures her.

M. Debierne, the discoverer of actinium is the chef du laboratoire & has most of the burden of the students. He is a Frenchman of the most charming type, gentle, kind, & courteous in manner, & with a vast fund of patience which he certainly needs, for the advice offered in every difficulty is, "Demander à M. Debierne." He knows a little English because I think he spent the summer in America, but he does not seem very proud of it for I have never heard it since the first day.[21]

In a later letter, Leslie adds the comment:

I trust my work will be in a fit form for publication when I leave here. It is really suffering from too much radioactivity. A number of people seem to be employing radium emanation at present and my electroscope is disgraceably sensitive to the influence of anyone entering from the "salle active" so that I spend half my time in keeping dangerous people out & in airing the room. Formerly more care was taken to prevent the distribution of activity all over the laboratory, but as the foundations for a new Institute of Radioactivity for Mme Curie are now laid, all precautions seem to have been abandoned.[22]

Although Brooks never published any of her work at the Curie Institute, we know what it involved, for three subsequent articles from the institute cited her contribution. Two of these papers were written by Debierne, while the other was written by one of Curie's workers, Lucie Blanquies. Debierne must have had Brooks performing parallel experiments on the recoil of radioactive atoms using the radium decay series rather than the thorium series. Debierne notes: "Certain things show that the recoil phenomenon is real. Thus Miss Brooks has shown that a freshly activated film allows particles of radium B to escape. As radium B is not volatile at ordinary temperatures, one can only explain this by the initial ejection of radium B particles at the moment of destruction of radium A."[23]

In his other paper, Debierne reports a study of the coefficient of diffusion of the actinium emanation in air. This, too, was an extension of Brooks's previous work, the only difference being that it involved measuring the diffusion rate of radon gas produced from actinium rather than from radium decay. The paper opens with the comment "The first experiments on the diffusion of the emanation from radium were performed by M. Rutherford and Miss Brooks." After remarking on other measurements, Debierne describes a different method that he has devised for measuring the rate of diffusion. He notes that the radiation level, unexpectedly, instead of declining, initially increased. He continues: "Miss Brooks has recently re-examined this question, at the laboratory of Mme Curie (the results obtained by Miss Brooks have not so far been published), and she has found that under certain conditions one can make the anomaly disappear."

Debierne then reports on his measurement of the coefficient of diffusion, adding: "The only hypothesis that is assumed is that which considers that the intensity of activity [radiation] is proportional to the concentration of the emanation [radon gas]. I recall that this hypothesis has been verified during the direct experiences of others than myself, and repeated recently by Miss Brooks."[24]

Brooks must also have been involved in measuring the decay rate of actinium B. Actinium B (a radioactive isotope of lead) is the second decay product of the radon produced from actinium (see Appendix 1). Blanquies notes: "At this time, one must remark that the decay constant for actinium B is poorly known: Miss Brooks, using an electrolysis separation, has obtained 1.5 minutes in spite of the fact that Hahn and Lise Meitner using a heating method have found a value of 2.15 minutes."[25] Incidentally, Blanquies herself concludes that the value obtained depends upon the method used, and that the discrepancy results from the difficulty of measuring such rapid decays in radiation.

From comments in Rutherford's correspondence, Brooks must have continued to correspond with him, though none of the letters have survived. In early 1907, Rutherford was in the process of moving back to England, this time to take up the professorship at the University of Manchester. Rutherford wrote from McGill to the then–head of physics at Manchester, Professor Schuster,[26] concerning a fellowship for Brooks at Manchester:

I was much interested to receive your second letter relative to Miss Brooks as a possible candidate for the John Harling Fellowship. Miss Brooks is a very good friend of my wife and myself and I should be delighted if she got the fellowship. She's at present working with Mme Curie at the Sorbonne and wishes to work at research in Physics in England, if possible. She is a very able woman with an excellent knowledge of mathematical experimental Physics. J.J. Thomson informed me that she was the best woman researcher next to Mrs Sidgwick, he had had at the Cavendish. Miss Brooks has already a good knowledge of experimental work in ionization of gases and radioactivity and I think would do most excellent work if she were given the Scholarship. At the same time, I should prefer she obtained the fellowship on her merits, quite apart from personal considerations on my part. If you would prefer to advertise the position, I am sure she would be willing to become a candidate, but in any case I feel quite confident she has as strong claims as any possible candidate for the position. I am enclosing a separate recommendation of Miss Brooks for official purposes.

The accompanying letter of recommendation for the John Harling Fellowship summarizes her career and confirms Rutherford's high opinion of her abilities:

Miss Brooks has a most excellent knowledge of theoretical and experimental Physics and is unusually well qualified to undertake research. Her work on "Radioactivity" has been of great importance in the analysis of radioactive transformations and next to Mme Curie she is the most prominent woman physicist in the department of radioactivity. Miss Brooks is an original and careful worker with good experimental powers and I am confident that if appointed she would do most excellent research work in Physics. I can strongly recommend her claims to the John Harling Fellowship.[27]

The John Harling Fellowship in Physics was a prestigious award. It was founded in 1900 and was awarded by the University of Manchester Senate, having a value of £125. The fellow was required to devote the whole of his or her time to research in the university laboratories under the direction of the professor of physics. The fellow could, with the approval of the senate, give an occasional course of lectures or demonstrations, or assist in teaching at the university from time to time, but he or she could not hold any salaried office.[28] It is of note that Brooks kept a description of the award among her surviving possessions.

Brooks moved from Paris to London about the middle of May 1906 at the invitation of Mary Rutherford, the spouse of Ernest. It is not clear whether it was simply a coincidence that Gorky and Andreyeva were visiting London at the same time. The Gorkys were in England to participate in the Fifth (London) Congress of the Russian Social-Democratic Party (the RSDLP), which was being held at the Brother Socialist Church, Southgate Road, Islington, London, from 13 May to 1 June 1907.[29]

We do know that Brooks must have met up with Andreyeva again in London. Andreyeva's son, Uri, was seriously ill in St Petersburg and Andreyeva wished to visit him, but she risked imprisonment on any attempt at entry into Russia.[30] She wrote to Ladyzhnikov from London in May of 1907: "It may happen that I will go to Russia. I shall take Harriet's passport, there is nothing to worry about ... we shall stay here until the 27th. If he doesn't strengthen I'll go to Petersburg."[31] Andreyeva did not use the passport on that occasion, but she did on her illegal return to her homeland in November of 1912. During that journey, she sent a telegram to Burenin, using the name Harriet Brooks.

It is part of the Brooks oral family history that Harriet gave her passport to Andreyeva. Harriet's sister, Elizabeth, recounted that to explain the disappearance of the passport, Harriet told the British authorities that she had lost it; accepting the explanation, they gave her a replacement.[32]

From London, Brooks wrote to Marie Curie, turning down an offer of continued work at the Curie Institute and promising to write up her research results thus far and send them to Debierne.[33] In this letter, she mentions a delay in choosing the winner of the Harling Fellowship, probably due to the absence of Schuster. On 11 June 1907, Rutherford wrote to Eve from Manchester and mentioned: "I run down to London next week where my wife is currently located with her mother. Schuster has been away on the continent but returns at the end of this week. I haven't heard anything definite about the John Harling Fellowship & your sister-in-law but think it is alright. I will not know till Schuster returns."[34]

It was during this visit to London that Brooks broke the news to Rutherford that she was to be married. He wrote Schuster in an undated letter to inform him of the news. The address on the letter was the same as the one where Brooks was staying, that is, 23 Princes Square, Bayswater. The building is currently a hotel, and it may have been such in those days. "Miss Brooks has just informed me that she is engaged to be married to a Mr. Pitcher of Montreal – formerly one of my demonstrators. He is an old and persistent admirer and has come over to England for the purpose of persuading her to go back with him. They are to be married next month. While personally I am sorry not to have her in Manchester, such bolts from the blue are to be expected when ladies are in question."[35]

Once again, Brooks's life was to change drastically – this time to take her away completely from the world of radioactivity.

CHAPTER 10

THE EVENTFUL SUMMER OF 1907

Before turning to the details of Brooks's engagement and subsequent marriage, we should look at the career choices that were open to her as a woman physicist. There were only two real options: she could have continued as a research assistant with Rutherford at Manchester or with Debierne in Paris, or she could have looked for an academic position at another women's college in Canada or the United States.

As a single woman in her thirties, she must have experienced the societal pressure towards marriage as well as the encouragement of her friends in this direction. In the book on feminism by John and Prestonia Martin, it is remarked that "child rearing is the noblest work an intellectual woman can do."[1] Maltby at Barnard also counselled that one should not forgo marriage for a career.[2] Thomas points out the dilemma: "Women scholars have another and still more cruel handicap. They have spent half a lifetime in fitting themselves for their chosen work and then may be asked to choose between it and marriage."[3]

Frank Henry Pitcher, whom Brooks married, was born 21 December 1872, the eldest son of the Rev. Joel Tallman Pitcher and Lucy Robinson.[4] He was educated at Stanstead College, Quebec, and at Montreal High School and then attended McGill University, graduating with an MSC in 1897. From 1894 to 1897 he was a demonstrator in physics at McGill, and a lecturer from 1897 to

1899. It was during these years that he almost certainly met Brooks. It is very probable that he was a demonstrator and/or lecturer of hers during her undergraduate years. Pitcher co-authored *A Manual of Laboratory Physics*[5] and published one research paper, "The Effects of Temperature and of Circular Magnetization on Longitudinally Magnetized Iron Wire."[6] Pitcher left the academic world in 1899 to become an engineer with the Montreal Water and Power Company; he was promoted to chief engineer in 1901 and to general manager in 1903.

Pitcher sent Brooks an enormous number of letters and postcards from the end of 1906 to the summer of 1907, most of which she kept.[7] Several of the letters help confirm Brooks's location at different dates. In addition, they contain many comments that give an insight into the course of Pitcher's pursuit of Brooks and provide clues about Brooks's own feelings.

It would seem from comments in the correspondence that Brooks became reacquainted with Pitcher during the summer of 1906. This would have been about the middle of September, for Brooks's letter of 13 September to Dean Laura Gill of Barnard had been sent from a Montreal address. Charles Gordon, husband of Harriet's sister Edith, knew Pitcher, and it was most likely that a meeting occurred through the Gordons.

The first surviving letter was sent from Montreal, probably to Brooks's Capri address. It is hard to judge the state of their relationship, as he addresses Brooks extremely informally (for the period) as "My dear Harriet" yet signs himself as "F.H. Pitcher." "Last week I found where you were hiding and make bold now to write again. The last one I sent to Paris as you told me to and I suppose it is still waiting for you at 'the Poste Restante.'"

As she had given him a Paris address, Brooks must have already spent some time in Paris, probably during late November or early December. He continues: "I can imagine you having a very delightful time in your beautiful surroundings. It has always been clear to me that people in your position – you in particular – are the luckiest and best off in the world." It is not clear whether he is envious of Brooks as a peripatetic scholar or as a leisured companion of the Gorkys.

Edith Gordon and her family were planning to visit Italy, and it would seem that Harriet was expected to return to Canada with them, for he adds: "You will be very jolly with your sister and the boys. Mrs Gordon [Edith's mother-in-law] tells one that she expects you to join them on their way back from Egypt."[8]

The next letter, addressed to the Villa Blesus in Capri, opens with "Dear Harriet" and is signed "Frank." It acknowledges receipt of a letter from Brooks, but unfortunately none of the correspondence in the opposite direction has survived.

> This morning I came down in the car with Mrs Gordon who told me that you had met your sister and the family at Naples. Mrs Gordon told me that you or rather Charles and the family were staying until April. Wonder if you will come back with them or if you will accept Mrs Rutherford's invitation to stay in England for the summer or what, you uncertain individual will do. Hope you choose the first.
>
> ... No doubt you enjoyed the opera at Naples if you went. You carry a deal of luck about with you don't you?"

The mention of Brooks's rendezvous with Edith in Naples would correspond to Andreyeva's letter to Ladyzhnikov (see chapter 9), in which Andreyeva mentions that Brooks was in Naples but would shortly be going to Paris. It may have been Edith's stopover in Naples that caused Harriet to return to Italy in the first place.

Interestingly, Pitcher adds a postscript: "Not saying anything about our affair. Want to see you first. F."[9] This indicates the respective families knew nothing of the relationship.

Brooks must have told Pitcher that she was returning to work with Curie, for the next letter was addressed to 3 rue Soufflet, Paris. "Beyond hearing that you were working in Mdm Curries [*sic*] laboratory, I could not discover what you were up to. So I was very glad to get your letter and to know more of your doings." Pitcher then informs Brooks that he is planning to visit Europe on business and hopes to see her during his time in Paris. "I gather from your letter, that if you are not having a gay time, you are still enjoying the life there. But what are you doing in another laboratory? Hurry up and discover some quicker way of decomposing or transforming radium, then you can have your airship, or anything else you want."

From the tone of this letter, it would seem that Pitcher did not approve of Harriet's returning to laboratory work. He continues:

"You spoke of a scholarship at Queens. I suppose by this time you have decided whether to take it or not. Mrs Gordon told me Mrs Rutherford said it was yours for the picking up. What about going to Norway? I hope you do not do so until I see you but if you get the chance to go for July and August, do by all means. I know a chap who fishes there then and he says its ripping at that time."[10] This is the only occasion where reference is made to a scholarship at Queen's. Brooks was probably being considered for the Robert Waddell Tutorship in Physics or Natural Science, though the sparse records for this period at Queen's University do not mention Brooks.[11] The allusion to a visit to Norway is interesting, as Ellen Gleditsch, who was working with Curie during part of 1907, came from Norway. Brooks and Gleditsch could have met at the Curie Institute, and Gleditsch might have invited Brooks to visit her country.

Pitcher's letters tend to ramble from topic to topic, with comments on happenings in Montreal interspersed between enquiries to Brooks. Then, turning to more personal matters, he writes in this same letter:

> Your apparently disinterested wish regarding my matrimonial destiny does not seem to be any nearer realization than when I saw you last. I hope you will let me talk it over with you when I see you. I will then tell you anything you want to know about my former attitude.
>
> I was really keenly disappointed Harriet that you did not come home with the others, but as I expect to see you soon I feel better about it. You might let me know as soon as you get this what your movements are likely to be.[12]

At this stage of his pursuit, Brooks seems to lack interest in marriage. His comment about a "former attitude" might refer to some personal differences between them when Brooks was a student at McGill and he was her demonstrator.

Brooks must have moved to London, England, about the middle of May, for Pitcher's letter of 16 May (signed F.H.P.) was forwarded from 3 rue Soufflet to the Imperial Hotel, Southampton Rowe, London. It is this envelope that establishes an approximate date of Brooks's move from Paris to London. He describes the travel plans for his European visit, adding:

I see of course the reason in what you say about the people you should go about with, but if you go to Norway in June or July I would make every effort to spend some time with you there. Always providing of course that you wanted me about. As to your destiny I have something to say to you about that when I see you and hope it may meet your views. It certainly is about time that you and I found out what we were going to do with ourselves. Do not go into any kind of a family as tutor. Not good enough for you. But not knowing the circumstances I should perhaps say nothing.[13]

Brooks must have been considering her future at this stage. Pitcher was certainly correct that a tutoring position in a family would be a waste of her talents and extremely low-paying. One can speculate that the reference to the company that Brooks was keeping might have concerned Gorky and Andreyeva. To someone of Pitcher's upbringing, revolutionary Russians would not have been appropriate as close friends.

Frank Pitcher duly arrived in London, writing from the Carlton Hotel, Pall Mall, London, to Brooks's address of 23 Princes Street, Bayswater, London, which she kept for the rest of her stay. Brooks must have made some remark about him, for he comments that "I do not quite like my pigeon hole that you have put me in,"[14] an indication that Brooks had some strong reservations about Pitcher's character. The next letter opened "Dear Love"; it was sent on 15 June 1907, 12.30 AM, from the Imperial Hotel, Russell Square, London, to the Princes Street address.

You have given me tonight something to think over in your saying that you cannot trust me.

I do not know of course how I would act under every circumstance of life, but I do know how I have acted as far as I have gone and with that knowledge I am perfectly happy in my love for you and the desire that you should be my wife ... I have absolute faith in you and know you will do what is right. But dear you must not expect me to take your words tonight as final for I know you love me and I venture to hope that you really think me capable of being worthy of you.

Now, I shall not give you up at all nor, unless absolutely prevented cease to see as much as possible of you. I must make good your trust of me but tonight I am quite in the dark as to its foundation unless my apparent indifference to you in the old days may be the foundation of your distrust.[15]

It is this second letter of Pitcher's from London that indicates Mary Rutherford was also staying there: "I shall see you tomorrow possibly before I go out to get you and Mrs Rutherford." In fact, according to one of Pitcher's earlier letters, it was at Mary Rutherford's invitation that Brooks had gone to England in the first place: "Wonder if ... you will accept Mrs Rutherford's invitation to stay in England for the summer."[16] The subsequent letter was sent in the early hours of 19 June (12.30 AM), announcing that he would be going to Europe for about a week. The letter indicates a turning-point in Brooks's attitude to Pitcher. He starts by expressing the right to manage her financial affairs. Brooks must have been relying on her family for financial support up to this time, and quite possibly the affluent Gordons had been helping out, as the Brookses were a close family. To Pitcher, the shift of her financial dependence to his shoulders would mark her firm commitment to a future with him. "I want you to get the things you need for yourself and it does not matter whether you have a remittance from home or not. I claim now the privilege of attending to these things. You may blow in now up to $250.00 which I shall be only too happy to see you do and settle for in a fortnight when I come back."

There is the suggestion in this same letter that Brooks is concerned that he drinks to excess:

You are altogether the most satisfactory and charming girl I have ever known. (NB I have not had a scotch tonight.) ... Since I have left the form and manner of our marriage in your hands I must ask you for a programme. The only thing I wish on is that it come off soon and before I must sail for home.

I have nothing in my heart Harriet but love for you. I trust Mrs Rutherford is not too mad to listen kindly to what you have to tell her tomorrow morning.[17]

It seems obvious that Pitcher had proposed and that Brooks would be breaking the news of her acceptance to Mary Rutherford. His expression of concern was possibly due to the vigour with which Ernest Rutherford had pushed Brooks's case for the Harling Fellowship; to have the candidate withdraw would have been somewhat embarrassing (the letter from Rutherford to Schuster indicates the degree of his surprise).

Pitcher then left for Europe. During his travels, he sent Brooks a large number of letters and postcards containing personal comments intermingled with commentaries on the scenery. Pitcher was particularly taken by the beauty of Switzerland, and he tried to extend his stay there as long as possible. He must have left the responsibility for the marriage arrangements completely to Brooks, for in a letter from Zurich, he comments:

> I do not like dear being away so long but I probably shall not have as good an opportunity of seeing the mountains I have so long wanted to see for some time.
>
> Trust you are working out a workable scheme for our marriage. I want to have it as soon as you hear from home. Because I must be getting back as soon as possible. Not that I am any good there but I feel I should be fussing at something useful.[18]

In the next letter, this time from Lucerne, he writes:

> I think it would be a decent thing for me to write Edith. What do you think? Or would you rather have me address myself to your mother?
>
> I trust you have not another bad time and think better – or worse – of your decision. Do spare me Harriet another siege of your love. Though I would gladly undertake it if necessary. You are a dear little solitary with ideas. Stick to them even if this boy is amused.
>
> What little fools you and I have been for some time past? We found this thing out last Autumn and by now – well of course we might both be dead; or less happy than we are.[19]

The comment on Brooks's having "a bad time" might indicate she was still ambivalent about her future. Possibly she was suffering from depression, and this was the point at which she had to commit herself irrevocably to a particular future. And her choices were poles apart. Pitcher's somewhat patronizing acknowledgment of Brooks's strong opinions is noteworthy. Having mingled with Fabian socialists and Russian intellectuals and revolutionaries, Brooks had been exposed to and had adopted views of life that must have seemed strange (or even bizarre) to someone like Pitcher, who had come from an upper-class Montreal background.

His reference to the previous autumn would reinforce the view that their relationship started when Brooks visited Montreal after leaving Summerbrook and before travelling to New York to rejoin the Gorkys for the transatlantic voyage.

Pitcher spent some time at Interlaken, sending daily letters as well as numerous postcards. In one letter, he mentions his plan to visit Paris on his way back to London and adds: "I wonder if you have written everybody telling them you are about to change your state. About Saturday they will be setting up at home about it and I shall have lost a lot I know of."

Before he finished writing this letter, he received one from Brooks. Hers must have been quite positive about the impending marriage, for Pitcher comments: "Your letter came all right and I got it after dinner. It was by far the most satisfactory I have read from you. Quite right you are in what you say about your present frame of mind, mine is identical but then I have the same mind for so long. Poor you, busy clearing up your affairs and chucking one thing after the other with regrets. But dear I shall try to make up to you all and more than you think you are now losing."[20]

Pitcher reached Paris on 29 June 1907, and on the thirtieth, he wrote a long letter after receiving a telegram from Brooks. The telegram must have shown what, according to Pitcher, was an uncharacteristic warmth: "I always thought you a cold sort of being my love until I read your letter of Thursday night. I have felt the same for you much longer and now how happy you make me in letting me see that you feel it too ... Think Prof Rutherford rather put his foot in it with his congratulations. If there are any flowers going about they should be thrown at this boy."

Obviously he felt that the congratulations should be in his direction as tribute to his successful pursuit of the hand of Brooks. He had overcome her considerable resistance to the idea, as shown in a later comment in this letter: "You seem to think lightly dear heart of my accomplishment with you, but I want you to know it was honest hard work, very much worth while. I did not think it a bit 'funny' at the time. You put me hard up to it all along but I wanted you so much I would not think of anything short of success."

During his stay in Paris, Pitcher visited some of Brooks's old haunts. In this same letter, he comments: "I love the Louvre. I shall go again on Tuesday after visiting your quarters across the river. Saw the Pantheon from a window in the Louvre and thought of your haunt close by at 3 Soufflet."

The letter suggests that they had originally considered marrying first and then telling the family later. However, as the following excerpt makes clear, Brooks had written to her mother, and Pitcher to Edith (Brooks) Gordon. The Mrs Eve mentioned was Harriet's other sister, Elizabeth (Brooks) Eve.

You are a very sensible girl Harriet to write your mother as you indicated to me. The other way is quite beneath both of us. If our affair is not good in itself it is not good for anything. If you like we can go together after we marry to see Mrs Eve.

I do not know whether I told you that I had written Edith. I asked her or rather hoped she would think well of our marriage, but at the same time I tried to leave the impression that it would come off anyway. I like your sister extremely and always have. She and I will be very good friends I am quite sure. You know dear girl I think, that Edith will never have cause to reproach me on the score of my conduct towards you – for at least want of trying. If I have a virtue it is in sticking up for those who have placed their confidence in me. Though you have to take my word for it now; but the fearless way you let yourself go has made me very proud indeed. That kind is all right.

Brooks seems to have been in favour of a civil marriage ceremony, while Pitcher argues (successfully) that in spite of her personal views, they should have a church wedding to satisfy the family. From his comments in this letter, Mary Rutherford would appear to have had a hand in the decision:

Mrs Rutherford has written me a very kind note congratulating me and telling me how happy you are. You cannot know how glad I am to hear that from other people. She says she has been "meddling" as to the form of our wedding. You probably know the rest. Now dearest as I have said that matter is in your hands and I do not care a button for myself one way or another, as long as we are legally married and I shall tell her so. But do you not think that both your people and my people would be better pleased if we were churched? It is as you know equally legal and when we can please others without any real harm to ourselves why not do it?

It is all very well to argue as I know you will about the inconsistency etc; but you are now coming to a point where you cannot live to yourself entirely and others must be considered. Life is a continual giving and taking. This is a place to give. Not that I care a fig dear as you know. But I have not your scruples and ideas about conformability and therefore it seems easy to me. I will not urge you beyond what you see above. All I want and all that matters is that we two should be married and go through the rest of our life together.[21]

Pitcher stayed on in Paris, mainly shopping and sightseeing with his business colleague, Will White. In his next letter to Brooks, it would appear that he had received a letter in which Brooks regretted their plans for a hasty marriage. This may have been prompted by a telegram from the Gordons (Charlie and Edith) to the Rutherfords, and Pitcher wondered why they had not written to Brooks herself. Possibly the telegram asked the Rutherfords to try and persuade Brooks not to rush into marriage so precipitously, particularly while she was so far away from the family.

Your extremely amusing letter of 9 PM Saturday is here. Why should you think Charlie and Edith would cable the Rutherford's rather than you if they knew your whereabouts? In any case it does not matter we would of course like all the blessings they have to offer but if they are not forthcoming we can get along very well thank you with just our own approval of one another.

Why you have been rash, I cann't [*sic*] see. You love me and I return your love. You know me better I think than you have ever known any man and yet you can trust me as you do and put all yours eggs in this one little basket. Altogether Harriet love it is a funny letter but cheer up I'll stick to you do not let things like that worry you.

I am also very much amused at the thought that in spite of all your clever scheming to avoid family witness at our marriage we shall at least probably have one; perhaps both Charlie and Edith are on the boat with all your family? Still better thought, perhaps they are bringing my pater over to see that we are properly tied up and sent down with good advice! Ah. Lordy I should like to see your face if all I can imagine might happen, came off. I know what you wanted when you wrote that letter, just me – conceited of course – that's all and then you would I know be amused instead of worried.[22]

This was the last letter that Pitcher wrote in his courtship, at least that survived, but on 5 July, Pitcher sent the only postcard with an actual message. The reason for her visit to Hull in northeast England is unknown. "Dearest H – Let me know how long you expect to be in Hull etc. Why on earth do I not get letters from you? Or are you too a busy! man. Saluted 3 rue Soufflet good and plenty yesterday. Having bully time F."[23]

News of the impending marriage must have travelled quickly around the family, for A. Stewart Eve at McGill wrote Ernest Rutherford on 7 or 8 July 1907: "We are glad to think that we shall soon see Harriet and that she will live in Montreal. I am looking forward to making Pitcher's acquaintance."[24]

Harriet Brooks was married to Frank Pitcher on 13 July 1907 at the Parish Church of St. Matthew, Bayswater, London. The witnesses were Charles B. Gordon, Harriet's brother-in-law, and W.J. White, Frank's business associate.[25] It is probable that the Rutherfords attended the ceremony. Harriet Brooks's marriage marked the end of her involvement with research, though her life continued to be eventful.

CHAPTER 11

ADAPTING TO MARRIED LIFE

After marriage, the Pitchers returned to Canada. Initially, they lived at 247 Bishop Street, Frank Pitcher's address prior to their marriage, then in 1909 they moved to 990 Queen Mary Road, where they lived for the rest of their lives.[1] In a letter of 1908 from Eve to Rutherford, he mentions the Pitchers: "The Pitchers are very happy and are going to build a hut as a holiday resort near the Chapleau Club in the Laurentians."[2]

Harriet Brooks never did pursue her research after marriage (as far as we are aware), notwithstanding her belief in a woman's right to continue a career after marriage (as explicated in her Barnard letters) and her own personal wish to do so (as implied in the letters of Frank Pitcher). However, the accepted philosophy of the day was that marriage and a scientific career were simply incompatible and that a professional woman had to choose between the two.[3] The one permissible exception to this rule was the woman who pursued research in partnership with her husband.[4] Another possible factor in Brooks's decision was that after Ernest Rutherford's move to Manchester University, there was no mentor to encourage her or guide her in any research activities. In fact, the whole nuclear research effort in Montreal had been in decline since the departure of its leader and his entourage.[5]

Brooks saved a number of letters from 1907 and 1908, though there is no indication why she chose to retain correspondence just from this period of her life. Four items were from Mary Rutherford, mostly describing life at Manchester, her attitudes to living in Britain, and some quite cutting remarks on "Ern's" (Ernest Rutherford) colleagues and adversaries. Mary Rutherford usually addressed Brooks as "My Dear Harrie," and in the first letter, from Manchester, she asked if the Pitchers were planning to visit England at Christmas. She enquired about Harriet's experiences with housekeeping and commented that she found it difficult to realize that Harriet was married. She must have been aware of Harriet's Russian friends, for in one letter she notes that she is enclosing a card from Russia that "Laura B. sent on to me."[6] One might speculate whether the card was simply a scenic postcard or a communication from Andreyeva, smuggled out of the country. The latter is certainly possible, especially as the cryptic sentence is quite unlike the more flowery prose of Mary Rutherford.

The second letter was sent from Edinburgh, where Mary had been visiting her mother, who was severely ill. After a long description of her rush to Scotland to visit her mother, she adds: "I was very interested in your camping trips and am amused over your housekeeping which you don't take very seriously. You need to be well off to run to hotels for dinners and even when you have it you may not always be able to run off."[7] Mary closes with the comment: "Kind regards to Mr Pitcher. Mrs Gordon evidently thinks I am responsible for him!" Mary Rutherford would appear from biographical sources to have been a very traditional woman.[8] She probably tried to convince Brooks that her future lay with marriage and children rather than with scientific research.

Mary Rutherford's third letter was sent from St Annes on Sea, Lancashire, but it only contained descriptions of various friends and acquaintances and a plea for information about life back in Montreal.[9]

The next letter came from North Wales, and in it Mary notes that "I heard him giving your name for the list of papers to be sent out." This indicates that Ernest Rutherford must have been keeping Brooks informed about his work. Brooks must have confided in Mary that she did not get very involved in the running of the household, for Mary remarks: "Your household seems to run wonderfully smoothly and with very little help from you! Esther must be a treasure your life seems one long picnic."[10]

The final letter, also from North Wales, again invites Harriet for a visit. "I think after such a hot summer you must need a change and ought to come across. Mind if you do you must stay with us we should love to have you and while Mr Pitcher was busy testing pumps we could have a grand old gossip unless you have grown superior to it. There has been such an impersonal air about your letters of late that I began to wonder if you felt it immoral to talk of people any more."[11]

The style of Mary Rutherford's letters is so different from the more intellectual type of communication in which Brooks generally indulged. It is difficult to envisage Brooks taking pleasure in writing the personal gossip that Mary seemed to enjoy.

Brooks also saved two letters that she had received from Prestonia Martin. Both of the letters were written from Grymes Hill, Staten Island, the Martins' winter residence. In the first (dated 28 October 1908), Martin refers to her concern for Marusia, a familiar name for Maria Andreyeva:[12]

> I am always glad to see your handwriting but was especially delighted this time because I somehow hoped you would have had news of Marusia. What *do* you suppose is the matter? I am beginning to feel sure that something has happened either she has gone back to Russia and doesn't dare write or she is ill. Surely your letter announcing your intention of going abroad would have brought immediate response if all were going right with them! What you told me this summer of her precarious position has increased my anxiety for her. Poor dear child! I am afraid she is doomed to sorrow she who was fashioned to give so much pleasure.[13]

The comment about Brooks's plans to go abroad, together with Mary Rutherford's invitation to visit in late 1908, might suggest that Brooks did indeed visit Europe and attend the Cavendish dinner of that year. It was at this dinner that the α-ray song was first performed (see Appendix 2).

The letter's reference to "what you told me this summer" would suggest that Prestonia and Harriet had recently seen one another. In support of this possibility, Brooks's memorabilia contains the following printed invitation to a summer school at Summerbrook:

> Mr. and Mrs. Martin take this method of announcing to their friends that they wish this year to assemble for the month of August a house party of fifteen to twenty persons, who will constitute a class for the

study of Fabian Socialism, under the direction of Mr. Martin. Mr. Martin will lecture for an hour every morning, using as a text book his own work on "Socialism Americanized" and explaining the Fabian attitude in relation to public questions. The afternoons will be devoted to outdoor enjoyments, and the evenings to music and informal discussion. Mr. Martin will be glad to direct the reading and farther the study of those desiring to inform themselves more fully upon his subject.

Prestonia Martin's second letter starts by discussing their mutual friends Maria Andreyeva and Nikolai Burenin (Eugenische) and refers to a letter received from Burenin. In view of Brooks's friendship with Burenin, it is probable that Burenin's letter had been sent initially to Brooks (perhaps via Laura B. and Mary Rutherford), then enclosed with her letter to Prestonia Martin.

I was so pleased to see this letter of Eugenitsch and get even this bit of cheery news from the dear noble boy who has tried so hard to do his part for the country and has suffered so cruelly and also so ineffectually. To think of the blood and tears shed in vain in the lost cause is enough to drive patriots to despair. It is a good thing that E is resolved to effect a "liquidation generale de tout nous passé" – There is nothing else for him to do and if he can get a measure of health back he will be happy in his music.

Do let me know what news you get in response to your letter. I cannot bear to think of what poor Marusia may be suffering. Her proud, tender, bruised heart will utter no complaint but I feel that she cannot endure very much more. How I wish we were near that we could get and comfort her a little![14]

At this time, Burenin was still in the thick of the Russian revolutionary movement.[15] The reference to Andreyeva might have concerned the continued illness of her son in Russia. She, too, was still active with the Bolsheviks and was threatened with legal prosecution for her participation in the events of the first Russian Revolution of 1905–1907.[16]

Brooks must have started looking for some activity outside of the home, for Martin responds to a query about possible philanthropic ventures: "Your enquiry about philanthropy (what are you going into now?) elicits from John the response that the best authority he knows is Mr. Devine head of our Charity Organization Society who is one of John's heroes. He has written a book on organized charity or something of the sort."

Martin then refers to a book she is reading – *Together* by Robert Herrick – in which, according to Martin, he diagnoses the problems of the leisure-class American woman. Martin disagrees with his conclusions and states that in her view the problem is that "she feeds her mind upon stimulants and condiments instead of upon food and doesn't take care of her health." Martin concludes, "Would there were more women like you, my Harriet!" and closes, "Ever Thine, Prestonia."[17]

Brooks continued to keep in contact with both the Rutherfords and the Martins, though no further letters survive. Both families paid periodic visits to the Pitchers' home.[18] She also must have continued to write to her Russian friends, but these letters too have not survived. As well, most of Andreyeva's archives were destroyed during the two world wars. However, in a letter she wrote to Burenin, sent from Florence in November 1907, Andreyeva comments: "Harriet is very worried about her own 'pauvre cher enfant' – ever since she married, she has behaved like a general grandmother [*babushka*] and treats us with tender patronizing."[19] We have no idea how long Brooks continued to communicate with Andreyeva, but in the memoirs of Andreyeva's son, Uri Zhelyabuzhskiy, Brooks is noted as his mother's closest friend.[20]

Although Brooks certainly seemed content, one wonders if the life of a "lady of leisure" in the upper part of Montreal society did not lack challenges for such an intelligent woman. Solomon contrasts two images of the married woman academic:

Ten years out of Wellesley in 1912, one graduate described her contented family with a professional lawyer-husband and two children. To the readers of *Women's Home Companion*, she presented a picture of the model wife, graciously entertaining her husband's friends, knowledgeable about public affairs, and busily engaged in social activities. Yet not all alumnae sounded as contented in class reports. Others, particularly those with small children, expressed guilt that they were not doing enough with their education. Elsie Frederickson, a Smith graduate of 1912, said: "And while I am ready to admit that I have an awfully good time with my nice husband and my little house and my silver and my funny daughters, I feel like a hopeless slacker all the time."[21]

Perhaps in response to this need, Brooks became an active member of Montreal's social organizations, where educated women tended to congregate. As Solomon notes, membership in such organizations provided academic women with an acceptable way to expand their interests and involve themselves in current social, cultural, or political issues. Among the societies that Brooks joined was the McGill Alumnae Society. At its meeting of 1 April 1910, she gave an account of the work of Marie Curie (the full transcript can be found in Appendix 3). The minutes of the society note: "Mrs Frank Pitcher followed with a short account of the work of Mme Curie, given with the clearness and life of personal knowledge."[22]

Brooks's opening remarks give some indication of her attitudes to women in physics and to her own abilities.

Perhaps it was a feeling of delicacy that prompted the committee to confine my attention to Mme Curie this p.m. that I might not be forced to expose the poverty of women's contributions to physical science. Even pure mathematics with such names as Mme Kovalevsky[23] and Miss Scott of Bryn Mawr[24] can make a braver showing. The combination of the ability to think in mathematical formulas and to manipulate skillfully the whimsical instruments of a physical laboratory a combination necessary to attain eminence in physics is apparently one seldom met with in women. I may seem to run counter to generally accepted views in this but it has been my experience that men are as a rule much more skillful in the manipulation of delicate and intricate work than women. It is no rare thing to see a man with hands twice the size of one's own and every appearance of being very dangerous to fragile furniture who will handle quartz fibres so fine that they can be seen only against a black cloth, with the same ease with which we would disentangle a skein of wool.

Mrs Mary Somerville is probably the only other woman whose work can be classed with that of Mme Curie.[25] She enjoyed the distinction according to Laplace of being the only woman who understood his "Mechanique Celeste" and her "Connection of the Physical Sciences" tho' written over half a century ago can still be read with pleasure and profit. That she was not deficient in experimental skill either is witnessed by the fact that she observed the magnetising effect of light rays, an effect that even with the more refined instruments of today is still too small for accurate measurement.

But I must not blame entirely the difficulties besetting the higher reaches of mathematics and our lack of deftness of hand for our neglect of the physical sciences. Their want of human interest has I think had much to do with our indifference to them. But the close of the old and the beginning of the new century has been marked by the astonishingly rapid development of a subject which has added a new philosophical interest to the phenomena of heat, light and electricity for our knowledge of radio-activity is bridging over the gulf separating matter and electricity.[26]

It is curious that Brooks shows such a negative view towards the role of women in the physical sciences. She is certainly correct in noting that there were few women who gained international recognition for their work; however, she makes no mention of some of the other women physicists of the late nineteenth century, such as Hertha Ayrton, Margaret Maltby, and Eleanore Sidgwick.[27] And she overlooks all the women of her generation, several of whom she must have known.

Brooks continued to be active with the alumnae society, giving a presentation in 1915 entitled "My Experiences with the Montessori Method."[28] Then in 1922, she hosted a large garden party in honour of the president of the Canadian Federation of University Women.[29]

Brooks was a charter (founder) member of the University Women's Club of Montreal,[30] and she was also active in the Women's Canadian Club (WCC). The Canadian clubs have two main purposes: to foster throughout Canada an interest in public affairs and to cultivate an attachment to Canadian institutions.[31] Members meet at intervals to hear distinguished Canadians or visitors from other countries speak on issues of national or international importance. The Canadian Club movement started in 1893, but it was not until 1907 that the first Women's Canadian clubs appeared, founded first in Montreal and Winnipeg.

The commonly held views of the members of the Women's Canadian Club tended towards the conservative end of the spectrum. One wonders how comfortable Brooks was in this milieu, given her exposure to radical causes through her friendship with the Gorkys. We can only speculate about the nature of her own views and philosophies. Certainly her preference for a civil marriage suggests that her ideas had been influenced by her stay at Summerbrook. However, her acquired status in the upper levels of Montreal society would have made it very difficult for her to participate in more activist organizations, such as the Montreal Local Council of Women.[32]

Brooks must have joined the WCC of Montreal almost immediately after its formation, for she is first listed as a member for the 1907–1908 period.[33] She served as honorary secretary in 1909–10, a post she held again for the 1911–12 period, and she was elected to the post of president of this organization in 1923.[34] The election was held at the annual meeting of the WCC in the ballroom of the Ritz-Carlton Hotel, and the venue would indicate the presence of a considerable number of members.

Because the president had a major role in the selection of speakers, we can gain an indication of Brooks's interests from the speakers chosen during her tenure of office and from the introductions that she gave.

These speakers included Count Albert Apponyi on the subject of the hopes for the League of Nations;[35] Princess Santa Borghese, PhD, on "Italy's New Literature and Theatre";[36] Thomas Whitney Surette on "The Art of Music"; Florence Kelly, general secretary of the U.S. National Consumers' League, on "Child Labor Legislation and the United States Courts"; and Rabbi Hillel Silver of the Jewish Reformed Church on "Christian and Jew; Will They Ever Meet."[37]

The final speaker of the year was Virginia Gildersleeve, dean of Barnard College, who was an interesting choice, considering Brooks's own sojourn there. Gildersleeve was to become one of Barnard's most famous deans.[38] Her presentation dealt with the history and role of the International Federation of University Women,[39] of which Brooks was also a member.

In her introductions to the speakers, Brooks made clear her own quite liberal beliefs. She felt that the post–First World War era, though hampered by "old traditions and old racial prejudices," had led to "new conceptions of liberty." In introducing Kelly, she commented on the progress that had been made in labour legislation. Her devotion to education came through in several of her opening remarks; she was interested, for example, in the "necessity of justly and wisely dealing with the matter of education [for Jewish children]." Finally, the breadth of her interests was indicated by her comment that music was the "highest expression of the culture of Western Civilization."

According to the obituary of Brooks in the Montreal *Gazette*, she was also active in gardening circles.[40] However, we were unable to find any details of her activities in this sphere.

In this chapter, we have concentrated on Brooks's correspondence with friends and on her intellectual activities after marriage, but there was another side – that encompassing wife and mother, the subject of the next chapter.

CHAPTER 12

FAMILY LIFE: FROM JOY TO SORROW

The Pitchers had three children: Barbara Anne Pitcher, born 19 October 1910; Charles Roger Pitcher, born 17 January 1912; and Paul Brooks Pitcher, born 5 August 1913.[1] Ernest Rutherford visited the family in April 1914, and he wrote back to his wife that the children were "all admirable specimens."[2] The Rutherfords visited the Pitchers whenever they were in Montreal, and in a letter to Bertram Boltwood, Rutherford commented: "Just a line to say we have arrived safely in Montreal where we will stay for a few days. We are all well & fit. I am staying at the University Club while my wife & Eileen [their daughter] are staying with Mrs. Pitcher (née Miss Brooks)."[3]

The Pitchers lived next to Sir Charles Gordon and Edith (Brooks) Gordon on Queen Mary Road.[4] The Pitcher lot was four acres in extent, while that of the Gordons was five acres. The two properties were run as a single unit with no fences in between. The Gordons had two tennis courts and a stable with horses, and both properties had extensive gardens. Harriet Brooks's son Paul recalls life at the house:

Tennis played a prominent part in activities, especially on weekends. On Sundays there were large tennis parties, interspersed by lavish luncheons and teas. I can recall HBP [Harriet Brooks] playing in the early years (in those ridiculous costumes women wore for sports in those days), but

I don't think she had any real interest in sports. Besides, much of her time was necessarily devoted to running the house for the benefit of her family and guests.

Her other interests were varied. She was a voracious reader in French and English and, to a minor extent, in Russian. Horticulture and botany were major interests (I have been able to salvage a number of valuable botanical books, which have been gratefully received by the local botanical gardens), and she kept in touch with the scientific world through visits from former colleagues, such as Ernest Rutherford. Uncle Stewart [Eve] and his family were, of course, frequent visitors, so she kept in close touch with both sisters, Betty and Edith.

Harriet kept up her friendship with Prestonia Martin and Mary Rutherford, but these two individuals lived far away and thus in later life her closest friend was Mabel King. King, the daughter of an English-speaking father and a mother of Huguenot extraction, was raised in a quiet, scholarly family in the French part of Montreal. She was a skilled linguist and had obtained a BA from McGill in 1907 and an MA in 1910. Paul Brooks Pitcher remembers King as having been witty and intelligent. Being about ten years younger than Harriet, she was a great companion to the children.

The Pitchers continued to travel after they married, regularly visiting England, Scotland, Switzerland, France, and Italy. Another of Harriet's sisters, Susan, a spinster, lived in Geneva, and they usually visited her during their journeys. Scotland was added to the ports of call after Sir Charles Gordon acquired an estate at Torridon on the northwest coast of Scotland.

The Pitchers had a contented marriage, Paul Brooks Pitcher recalls. Frank Pitcher had a "robust and energetic nature, sometimes boisterous and full of fun. What he loved was the backwoods and the companionship and activity of sporting life." Vacation highlights for the Pitcher family were the expeditions to a fishing camp on the St Lawrence River. Paul Brooks Pitcher notes:

Father's sporting diaries reveal that, as early as 1912, she [Harriet] joined a salmon fishing expedition on Uncle Charlie Gordon's yacht in the Lower St Lawrence, where they fished the Mingan, Corneille, and Thunder rivers and rivers on Anticosti Island. A few years later, he acquired, from the Mingan Seigniory, the lease of the Jupitagan and Magpie rivers, which were about seven miles apart on the north shore of the Gulf of St Lawrence, opposite the west point of Anticosti Island.

A camp was made by drawing together two abandoned fisherman's cottages, and an old fishing schooner was acquired for travel between the two rivers. As we spent from six weeks to two months there every year, the provisioning for the Jupitagan trip was a major feat of organization, in which both HBP and Father both took part. All the supplies had to be pre-ordered from Robin, Jones & Whitman in Quebec City and loaded on a steamer to accompany us on the trip to the river, which took two days or more. As fresh meat would not keep, I can recall bringing down two live lambs to be slaughtered later, as required (as we children became very fond of these animals, we kicked up an awful fuss and refused to eat for days when they were slaughtered). Another source of meat was Anticosti Island: Father would obtain permission from Meunier (the French chocolate king who then owned the whole island) to camp there and shoot deer for the pot. We would cross over in the old schooner, make a tented camp, then shoot and bring back as much deer as we could handle.

Here, again, there were numerous cousins, other relations, and guests. HBP fished occasionally, but unenthusiastically and, in her spare time, seemed to prefer searching for rare species of wild flowers and the company of Mabel King, who frequently accompanied the family on these trips.

In the 1920s, the Pitchers also purchased 1,100 acres in the Laurentians. The property included three small lakes and was situated about sixty miles north of Montreal and seven miles west of the railroad station at Val David. Reaching the property involved a horse and buggy ride in summer and horse and sleigh in winter. The home, which lacked electricity and telephone, was built overlooking one of the lakes and close to a farm that was also part of the property. Paul Brooks Pitcher describes the pleasant days at the country home:

In the spring and autumn (we were, of course, at the Jupitagan most of the summer), the activities consisted of trout fishing in the lakes and long walks in the mountains. HBP indulged in country gardening and I can remember the masses of hardy narcissi surrounding the house, which she had planted. In the winter there was snowshoeing and cross-country skiing, for which the terrain was ideally suited. The adult group would set out on snowshoes and our generation would set out much later to ski on the trails broken by them. Setting snare lines to obtain hares for the pot was also an occasional activity in winter. It served mainly as a weekend retreat and here again there was the usual number of relations and friends, as guests.

We have an account of life at all three residences from Cicely (Eve) Grinling, daughter of Elizabeth (Brooks) Eve:

I only knew Aunt Harriet well when I was aged five to twelve, that is, between 1920 and 1927. I was often sent to stay with her in the brown stucco house in the Côte des Neiges [part of] Montreal, in the Laurentian mountains at Ste Agathe, or summer holidays at the Jupitagan and Magpie salmon rivers in the St Lawrence River. The last were most memorable. A long boat journey from Quebec, there was no road or rail, the boat called once a month with provisions and there would be twelve to twenty people staying in the house. One French Canadian family lived near the house on the Jupitagan; the village on the Magpie was five miles away. We lived on cod, trout, or salmon and berries in season. Days were spent fishing, evenings with parlour games, anagrams, rhyming conundrums, bridge, chess, and Uncle Frank screeching painfully on the violin. Most of their guests were schoolmasters or university academics. And Aunt Harriet presided with calm amusement. She never shone but glowed with a steady warmth and unobtrusively looked after everyone's needs.

My mother could be too witty and clever, sometimes I felt at my expense, but Aunt Harriet was the rock of my childhood, never ruffled, always outwardly serene and calm. She also had the gift of making one feel, even as a child, that one was important and included in her life.

In the Laurentian mountains I still see her amazing collection of naturalized narcissi, and in Côte des Neiges a naturally beautiful garden in contrast to her sister Lady Gordon's more opulent greenhouse, conservatory, and bedding plants in the next-door property.[5]

Another account of the visits comes from Joan (Eve) Denny, the elder daughter of Elizabeth (Brooks) Eve:

Aunt Harriet was married the year I was born. I remember her as a warm, kindly person. She asked me out to stay for several days at a time to play with her children who were slightly younger than I was – probably when I was nine to fifteen years old – two huge gardens, each two acres or more, was a perfect place for four children – apple trees, vegetable garden, where we were allowed to eat whatever we liked – I think Aunt Harriet must have known that a tomato right off the plant was much tastier than those at the table. As I recall, she was always there – in the lovely big living-room (it was always sunny then) making beautiful smocks for my cousin Barbara. Having her friends to tea – we escaped as soon as we'd politely said, "How do you do" – I don't think I was particularly interested in grown-ups. [It was] more fun to play hide and seek, climb apple trees – this for me was between the ages of ten and fifteen. But I remember Aunt H as always being there – when I broke my arm – or to ask if we could "go down to the Greeks" for candy or ice-cream. She never seemed to get cross with us, or be upset or fussy about what we did – just about a perfect aunt, I thought.

I was up at Lac Normand a couple of times. Barbara and I (12–14 yrs?) were allowed to go and spend the night at our other aunt's cottage by ourselves – obviously she trusted us, and when grown-ups like Aunt H are wise enough to trust you, children usually behave accordingly. The visit to Uncle Frank's salmon river down the north shore of the St. Lawrence – five children: three Pitchers, my little sister, and me – was at bit boring for children – black flies, mosquitoes, and muskeg kept us mostly inside. Aunt Harriet must have found us a bit trying at times. But my memories of many visits, you can see, were happy ones, and looking back I can see that my aunt's warm and understanding nature made it so.[6]

Unfortunately, this idyllic life came to an end with a tragic series of deaths in the 1920s. Harriet's father died on 20 November 1920 and her mother on 14 April 1922. Then, less than four years later, on 20 January 1926, her son Charles Roger succumbed to spinal meningitis at Val David.

However, the most traumatic experience was most certainly the disappearance of her daughter, Barbara. This occurred on 21 March 1929. The story was described in the Montreal *Gazette*, the first mention being on 23 March:

Miss Barbara Pitcher, daughter of Mr. and Mrs. Frank Pitcher, of 990 St. Mary's Road, has been missing since Thursday morning, after being left by her Chauffeur at the Arts Building of McGill University a few minutes before 9 o'clock.

Her movements on that morning as far as have been ascertained, were as follows: She left her home at 8.50 and was driven to the Arts Building to attend a class in German, which she failed to do. Shortly after 9 she was seen going towards the Royal Victoria College along Sherbrooke street. It is thought probable that she went to the library there and worked. At 10 o'clock she was seen on the steps of the Arts Building. Since then trace of her has been lost.

Mrs. Pitcher said that her daughter had been feeling rather tired as a result of the work in which she had been engaged and had expressed discouragement with the progress she was making, but was very unwilling to give up.[7]

At the time, the Pitchers thought that Barbara had been kidnapped, given her relationship to the very affluent Gordons. They telephoned Herbert Brooks (brother of Harriet) to suggest that special precautions be taken with his son, Franklin, in case a further kidnapping attempt was made.[8]

A later newspaper report gave more details of the disappearance: "When she left her home that morning she had not taken her books and she had failed to don a pearl necklace and a wristwatch that she was in the habit of wearing. When at night she failed to return home her parents became alarmed and a search was commenced. It was known that she was in a fatigued condition and it was feared that she might be suffering from loss of memory."[9]

On 25 March a reward of $1,000 was offered for finding her,[10] and the next day students and police undertook a major search of the slopes of Mount Royal, as it was thought that Barbara might have wandered onto the mountain.[11] During the day, aircraft circled the city dropping leaflets carrying photographs and a description of the missing woman. The leaflets also announced an increase in the reward, to $5,000.

The story was mentioned again on 28 March,[12] when it was claimed that someone of Barbara Pitcher's description had been seen in the towns of Ste Rose and Ste Thérèse, but the individual proved not to be her. The police announced that they had been inundated with telephone calls, but that all the leads had been fruitless. The report of the following day again described the searches in progress involving the city police, McGill faculty, a large number of McGill undergraduates, and private detectives.[13]

It was not until 7 May that the tragic news was received that her body had been found in the water at the dam across the Rivière des Prairies at Sault aux Récollets, concluding the "long and continent-wide search." The case of the missing student must have been a major item of interest in Montreal at the time, for the article concluded with the comment: "The recovery of the body and the report of the medical men which showed that there were no marks of violence nor any sign of foul play will set at rest the persistent rumors that had been circulating during the last few weeks."[14]

The closing of the case came the following day with the coroner's report, which concluded: "Summing up the features of this strange case Coroner Prince saw no reason to suspect foul-play and so rendered his verdict of 'found drowned.'"[15]

As a final trauma, Paul Brooks Pitcher, the only surviving child, became seriously ill less than two years later. He comments: "Soon after commencing studies at McGill in September 1931, I contracted tuberculosis. This was another blow to both HBP and Father, as the life of their last child appeared threatened. However, a year "curing" at Saranac Lake, NY, solved the problem, and thereafter, I don't believe I gave them more trouble than the average spoiled adolescent youth."

Paul sums up his view of his mother as follows:

> I can only add that HBP was of lively temperament, placid without being phlegmatic. I cannot recall her ever being fussed or angry and she always answered her children's questions directly and without evasion.
>
> Her household duties in our various establishments were vast, but not physically taxing, as there always seemed to be an abundance of servants or other hired help in attendance. Nevertheless, these had to be hired, trained, organized, and instructed, and this is where her organizational skills came into play.

Harriet Brooks Pitcher herself died at the age of fifty-six on 17 April 1933. Paul ascribes his mother's death to radiation-related diseases: "Mother's death followed a lengthy illness which I can recall as having been described as 'blood poisoning' or a 'blood disorder.' It now occurs to me that it could well have been leukaemia, a legacy of her earlier exposure to radiation while handling radium during a period when protection had not been considered necessary."

During Brooks's years of research, radioactivity had been treated in a very cavalier fashion. The Curies in particular had noted the severe skin damage that exposure to radium could cause,[16] but they regarded this as simply an occupational hazard. Rutherford, too, did not appreciate the precautions needed to prevent contamination of the McGill laboratory. Subsequent radiation surveys found a great deal of radioactive contamination on floors, benches, and window-sills, which had to be cleaned up to render the building safe for occupation and suitable for measurements of low-level radiation.[17] Another illustration of the potency of the radiation sources relates to a visit Rutherford paid to Dartmouth College. During his demonstrations, Rutherford discarded the paper he had used to transfer some radium salt into a tube. This paper was saved by his host and used as a radioactive source for a period of forty years.[18]

Of particular importance, Brooks had worked on a routine basis with radon gas ("emanation"). It was in 1921 that the association between radon and lung cancer was first proposed, but the hypothesis was not accepted for another thirty years. We are now particularly aware of the hazards of radon.[19]

The atoms of radon decay to polonium, releasing α-rays in the process. The polonium decays to lead, and the lead to bismuth, each transformation again being accompanied by α-radiation. Polonium atoms formed in a room will stick to dust particles, and these particles will be inhaled and will adhere to the lung surface. As each decay occurs, the lung surface is radiated. In addition, some of the atoms, particularly lead, will be absorbed into the bloodstream, where they can irradiate other tissues. Thus, radon is of far greater long-term danger than radium or the other solid radioactive sources.

A. Stewart Eve must have written to Rutherford with the news of Harriet Brooks's death, for Rutherford sent a handwritten letter back to Eve on 6 May 1933. It was unusual for Rutherford to have written a letter by hand, as his handwriting had deteriorated to the extent that he usually typed his correspondence.

It was very good of you to write me news about Harriet Pitcher. I had not heard of her illness and a few days before your letter, Mary had spoken of writing to her. It is a very sad business. The last time she came to see us about two years ago, one could not but recognize the obvious loss of vitality but this was quite understandable after her family calamities. I have the happiest remembrances of our friendship in the old days at McGill and the renewal of these during our occasional visits to Montreal. She was a woman of great personal charm as well as of marked intellectual interests. I am afraid her domestic life was not without serious trials which she bore with astonishing fortitude. My wife and I held her in great affection and her premature death is a grievous blow to us. I shall see whether I can compose a short statement of her scientific contributions for "Nature" in the next few weeks.[20]

Rutherford did indeed write an obituary for *Nature*, giving his view of the important aspects of her work.

Harriet Brooks (Mrs. Frank Pitcher) who died in Montreal on April 17, was well known in the years 1901–5 for her original contributions to the then youthful science of radioactivity. Distinguished graduate of McGill University, she was one of the first research workers with Prof. (now Lord) Rutherford in Montreal. She observed that the decay of the active deposit of radium and actinium depended in a marked way on the time of exposure to the respective emanations and determined the curve of decay for very short exposures. This work which was done before the transformation theory of radioactive substances was put forward, assisted in unravelling the complex transformations which occur in these deposits. With Rutherford she determined the rate of diffusion of the radium emanation into air and other gases. These experiments were at the time of much significance, for they showed that the radium emanation diffused like a gas of heavy molecular weight – estimated to be at least 100.

Miss Brooks entered the Cavendish Laboratory, Cambridge, in 1903 and continued her radioactive investigations. In a letter to NATURE of July 21, 1904 (vol. 70, p. 270) she directed attention to a peculiar type of volatility shown by the active deposit of radium immediately after its removal from the emanation. In the light of later results of Hahn and Russ and Markower in 1909, it is clear that the effect was due to the recoil of radium B from the active surface accompanying the expulsion of an α-particle from radium A. This method of the separation of the elements by recoil ultimately proved of much importance in disentangling the complicated series of changes in the radioactive bodies.

After her marriage to Mr. Frank Pitcher of Montreal, she gave up her research work but took a strong interest in university affairs. In this she was aided by her family ties, for one sister is the wife of Sir Charles Gordon, a prominent supporter of McGill University, and another the wife of Prof. A.S. Eve, professor of physics in McGill University. A woman of much charm and ability, she was a welcome addition to any research laboratory and left in all who met her a vivid impression of a fine personality and character. R.[21]

The Montreal *Gazette* published an obituary with the heading "MRS. F. PITCHER DIES AT RESIDENCE: Noted Physicist Had Worked With Thomson, Rutherford and Mme. Curie: GRADUATE OF MCGILL: Gained Brilliant Reputation by Scientific Discovery – Greatly Interested in Horticulture." The obituary summarized her work in physics, though with considerable simplification: "The world of science has lost a distinguished figure and the city of Montreal a much loved personality in Harriet Brooks Pitcher, who passed away yesterday at the age of 56, following a lengthy illness. Discoverer of the recoil of a radioactive atom, her work in physics was widely recognized and gained her a brilliant reputation."[22] However, it does go on to describe her subsequent interests in horticulture: "To many people in Montreal Mrs Pitcher will be well remembered as a lover of literature and of flowers and gardens. During her distinguished career she was keenly interested in horticulture, and her flowers were always one of her greatest delights."

Part of the obituary in the *McGill News* describes her later life more fully:

After her marriage in 1907 to Frank H. Pitcher (Science 1894), Mrs. Pitcher gave up her active participation in scientific work, and turned her attention in other directions. In 1923–24 she was President of the Women's

Canadian Club, and for several years she served as a member of the Scholarship Committee of the Canadian Federation of University Women. Members of the McGill Alumnae Society will remember the delightful garden party given in honor of Dean Moyse in the grounds of her residence on Queen Mary Road.

Although Mrs. Pitcher's public activities were so noteworthy, it was in a more restricted and intimate circle that she was most admired and appreciated. Too often a woman of distinguished attainments pays the price of her scholarship by a loss of those feminine qualities which endear her to those in her immediate surroundings. It was not so with Mrs. Pitcher. The most gracious of hostesses, the kindest of friends, she was truly beloved by all who knew her. Her gentle and unassuming manner, her low pitched voice, her keen sense of humour, her never failing interest and sympathy, all contributed to the charm of her personality.

Mrs. Pitcher's garden was one of her keenest delights, and during many years in Montreal and at her country place at Val David, she proved the truth of Bacon's words, "Gardening is the purest of human pleasures and the greatest refreshment to the spirit of men." Her knowledge of plants, of trees, of birds, her joy in the beauties of forest, field, and sky, and in all the varied manifestations of nature, were well known to all her friends.

Not less remarkable was her knowledge of literature. Never led astray by the false or meretricious, her judgment was exceedingly sound. She was an example of the value of education in training the powers of the mind, in purifying the taste, and in developing that discrimination which enables us to distinguish the genuine from the spurious, the essential from the non-essential.[23]

To complete the unfortunate string of family deaths, Frank Pitcher died on 21 August 1935 of pneumonia, at the age of sixty-four.[24]

CHAPTER 13

WHY HAS BROOKS BEEN OVERLOOKED?

There are many reasons why we should not be surprised that the life of Brooks has been overlooked by historians. Although the discipline of the history of science originated in the 1920s and 1930s, it was not until the 1970s that the history of women in science became a popular field of study in the United States and Europe.[1] In Canada, this field is only now developing. Such lateness is due in part to the overwhelming demands on the few Canadian women historians.[2] But even more the fault lies with science historians in general, for it cannot be claimed that the work of women scientists went unnoticed by their contemporaries. In his texts, for example, Rutherford always acknowledged the work of all his students (including the women). One of Rutherford's later students, Peter Kapitza, remarks: "Rutherford was very particular to give credit for the exact authorship of any idea. He always did this in his lectures as well as in his published works. If anybody in the laboratory forgot to mention the author of the idea Rutherford always corrected him."[3]

In addition, science history has been looked at from the hero/heroine perspective;[4] that is, each new discovery has been associated with some "great name." As we have seen, in accounts of early nuclear science, the names of Rutherford, Thomson, and the Curies have always dominated the discussion. The structure of nuclear research essentially involved these three individuals and their research groups and associates. All the others, such as Brooks, were viewed as mere secondary players.[5] Ainley refers to these assistants who performed much of the actual work as the "invisible" scientists, many of whom were women.[6]

This study of Harriet Brooks highlights some aspects of the sociology of science. Very few women were able to make it into the big league (the "successful" stream); most professional women scientists were relegated to the "obstructed" stream, where their careers hardly advanced at all, and as in Brooks's case, terminated upon marriage.[7] The fact that she did not become one of these heroines in the Curie mould doomed her to obscurity in the eyes of the traditional historians of science.

However, she was not completely forgotten. Schofield gives a significant description of her contributions.[8] Also, the 1933 review by Mary Weeks of the discovery of the radioactive elements does mention Brooks's work and brings to light the role of some of the other women pioneers.[9]

The phenomenon of transferring credit to the famous researcher and away from the lesser-known person occurs irrespective of sex. For example, much of Debierne's work is identified as that of Marie Curie. This transference is known as the Matthew Effect, and it is defined by its originator as "the accruing of large increments of peer recognition to scientists of great repute for particular contributions in contrast to the minimizing or withholding of such recognition for scientists who have not yet made their mark."[10]

There is, however, one additional factor operating in the case of many women: instead of staying for extended periods in nuclear research, the female researchers became teachers (such as Fanny Cook Gates) or married and raised families (such as Brooks), unlike male scientists such as Soddy and Boltwood who spent their entire lives working in the same field. Brooks never stayed active in physics long enough to become listed in the common reference source, *American Men in Science*.[11]

Brooks is rarely mentioned in reviews of early women in science. Of four compendia of women scientists in history, three make no mention of her at all,[12] while the fourth notes, "Harriet Brooks, physicist (18??-19??)."[13] It is particularly regrettable that she is not included in current compendia of Canadian biographies;[14] yet in 1912, she was listed in *Canadian Men and Women of the Time* – a significant achievement![15] She is not even mentioned in Canadian high school physics texts.[16] This is not completely unexpected, for, as Arnold remarks, excellent women scientists have traditionally been ignored, while men scientists of the same calibre are remembered for their particular contribution:

> How great is great enough? Apparently a double standard exists. In order to be included in science curriculum materials, men can be super-great, great, or just successful scientists. But even successful women scientists who are recognized as having made a significant contribution in the eyes of their peers – those historical figures who have been starred in *American Men of Science* or contemporary contributors who have been elected to their respective professional societies or even to the National Academy of Sciences – are not included.[17]

A brief and partially erroneous outline of her life appears in the *McGill News*,[18] and in her study of women at McGill University, Gillett includes a reference to the accomplishments of Brooks as a student and in terms of her founding role at the Royal Victoria College.[19] Finally, a simplified description of her problems at Barnard College appears in Rossiter's work on women in academic science.[20] It is significant that the discussions of Brooks in the *McGill News* and in Rossiter's work are written from the perspective of "promise unfulfilled"; that is, the references describe less of what she accomplished and more of her abandonment of a promising career, though as we have discussed, opportunities for her further advancement were quite limited.

Rutherford remarked that, next to Curie, Brooks was the most outstanding woman in the field of radioactivity. He credited her identification of emanation (radon) as a vital piece of work that had led him to propose the theory of the transmutation of one element into another. This was a revolutionary view for a time when the concept of the immutability of atoms was a basic part of scientific dogma. A subsequent piece of her research led to the idea of the successive changes of elements, which Rutherford noted was a key step in untangling the complexities of radioactive decay. Finally, there was the discovery of the recoil of the radioactive atom – the item that has been most associated with Brooks's name.

One might first question whether Rutherford's comment was a rather hollow statement of praise, for were there that many women researching into radioactivity? In spite of the common myth to the contrary, there seem to have been numerous women publishing research work in nuclear science around the turn of the century. In fact, with the obstacles to advancement in the physical sciences, it is more surprising how many there were rather than how few. These pioneers included Winifred Moller Beilby, Lucie Blanquies, Fanny Cook Gates, Ellen Gleditsch, Ada Florence Hitchins, Stephanie Horovitz, Elizabeth Rebecca Laird, May Sybil Leslie, Lise Meitner, Ruth Pirret, Eva Ramstedt, Jessie Mabel Wilkins Slater, Jadwiga Szmidt, and Edith Gertrude Willcock,[21] all of whom published work in nuclear science – an appreciable proportion of the contemporary researchers.

The presence of a number of women in nuclear science should be expected, for as Rossiter points out, employment for women in "Big Science" was in the rapid-growth areas where demand for personnel resulted in the hiring of qualified women support staff.[22] Yet, at the same time, one gains the impression from the correspondence of some of these women pioneers that there was more to their involvement in nuclear science than simply job availability. In those early years, the field was exciting and new, and hence appealing to these ebullient women who were looking for a purpose in life. As Gornick describes in her interviews with modern women scientists: "Each of them had wanted to know how the physical world worked, and each of them had found that discovering how things worked through the exercise of her own mental powers gave her an intensity of pleasure and purpose, a sense of reality nothing else could match."[23]

The early days in nuclear science resemble those of astronomy. A significant number of women had flocked towards this new and exciting frontier of physical science as well.[24] Then, as both fields matured, the proportion of women dwindled. For example, in a photograph of the workers with Rutherford and Thomson in 1932, there are only two women (Miss Davies and Miss Sparshott) among the thirty-seven researchers.[25] The two fields are also similar in that women had difficulty penetrating the upper ranks of both.[26]

There are other specific reasons why Brooks has received so little credit. Because her first paper, which proposed that "emanation" was a new gaseous element, was published jointly with Rutherford, accounts of the research usually extol it as Rutherford's work (the Matthew Effect).[27] Again, there is a parallel with astronomy: while Jocelyn Bell discovered the first pulsar, a bizarre astronomical object, and, to a large extent, explained her findings, it was her supervisor, Anthony Hewish, who received the Nobel Prize for physics (1974) for this discovery.[28]

In many ways, it was Brooks's painstaking work on the decay products that broke the ground for the most important discovery – the successive changes of elements. The written account of Rutherford's Bakerian Lecture on this discovery was published under his name alone, and his frequent references to the work of "Miss Brooks" went unnoticed by subsequent historians of science.

Oddly enough, however, an additional problem was that much of her work was published under her name alone. Many historians of early radioactivity experiments would probably have overlooked her contribution as they searched the records of Rutherford, Thomson, Curie, and other well-known individuals.

As for her time at the Curie Institute, it is only from the publications of Debierne that we can obtain an appreciation of her research there. It is disturbing that the biographies of Curie make no mention of her research workers, as if their contributions would detract from Curie's own reputation. In fact, in Curie's autobiographical notes, the sole mention of any collaborators is with regard to her taking over the laboratory after the death of Pierre Curie: "A few scientists and students had already been admitted to work there with my husband and me. With their help, I was able to continue the course of research with good success."[29]

Also, as Rutherford noted in his obituary for Brooks, we should realize that the value of her work could not really be appreciated until the phenomenon of radioactivity was better understood. Her work was done before the theory of the transmutation of the elements was accepted – in a period when research workers were groping in the dark for explanations of these strange phenomena.

Her discovery of the recoil of the radioactive atom is a particular case of delayed recognition, that is, a discovery made before the framework of science could appreciate the significance of the result.[30] Added to that, it is not uncommon for unusual findings to be ignored if they come from a young researcher.[31]

As her biographical details have not been compiled until now, Brooks has been no more than a name on some research papers. Thus, science writers could only make passing reference to her,[32] and her name has never gained wide recognition. To compound the problem, few women (including Brooks) felt that their contribution was significant enough to warrant saving their papers and correspondence.[33] We are fortunate indeed that Brooks kept a few items from her life in the crucial 1907–1909 period, for otherwise other aspects of her life, such as her friendship with Prestonia Mann Martin and Maria Andreyeva, would never have surfaced. No one would ever have thought of looking in the memoirs of Russian revolutionaries for information on a physicist! It is most regrettable that Brooks never wrote an autobiography, for she was probably the only person to work with all three of the great physicists, Thomson, Rutherford, and Curie. As well, literary and political historians would have certainly been interested in her accounts of life with the Gorkys.

In spite of her own periods of self-doubt, it is apparent that Brooks was a talented experimental physicist who performed some valuable research in the early days of radioactivity and who thus deserves to be better known. Of importance from a historical perspective, her life – more typically than that of Marie Curie – reflects the experiences of a talented woman physicist of the time.[34] As such, her story provides a valuable case study for those who wish to understand and illuminate the role of women in science.

APPENDIX 1
THE RADIOACTIVE DECAY SERIES

It is important to realize that Brooks was working in the earliest days of the study of radioactivity. Not until 1902 was it first suggested that radioactivity involved the change of one element into another. Alchemy had long been discredited, and Rutherford and Soddy received much opposition to their theory of the transmutation of the elements. Even the Curies were initially opposed to the concept.

During Brooks's time, and for many years later, the different elements produced by radioactive decay were represented as some form of the starting material. Thus, when reading Brooks's reports, we have to "translate" what is meant by "Th Emanation" or "Radium B." To help the reader, the following three pages list the decay sequences for uranium, thorium, and actinium, the three elements with which Brooks worked. To be more precise, most of her studies were on the gaseous radon produced in the middle of each decay sequence and on the two products that followed from them.

For simplicity, only the major decay pathways have been represented. It should be noted that some additional alternative routes have been found in more recent times. For example, in the decay of uranium-238, not quite all of the bismuth-210 decays to polonium-210. Instead, 2×10^{-4} percent decays to thallium-206.

THE URANIUM SERIES

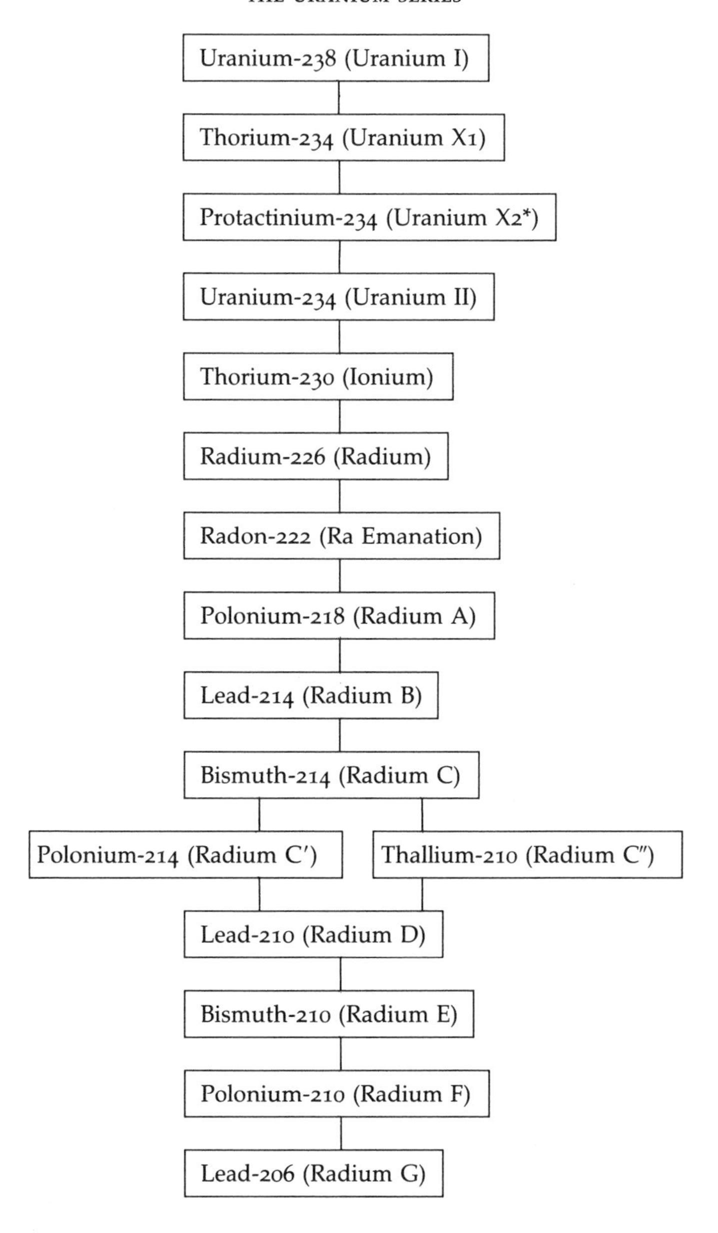

THE THORIUM SERIES

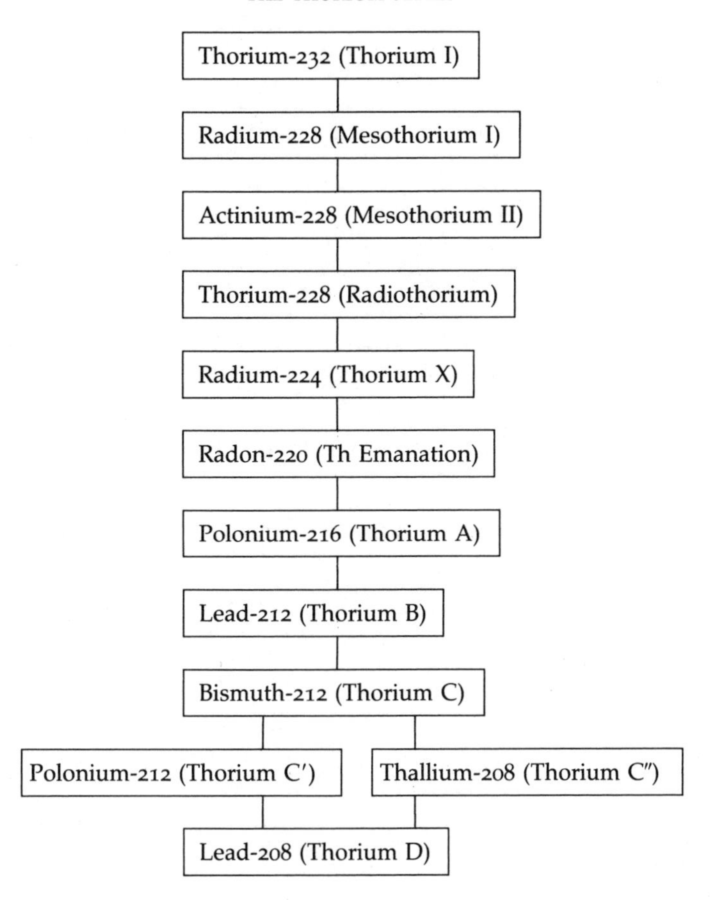

THE ACTINIUM SERIES

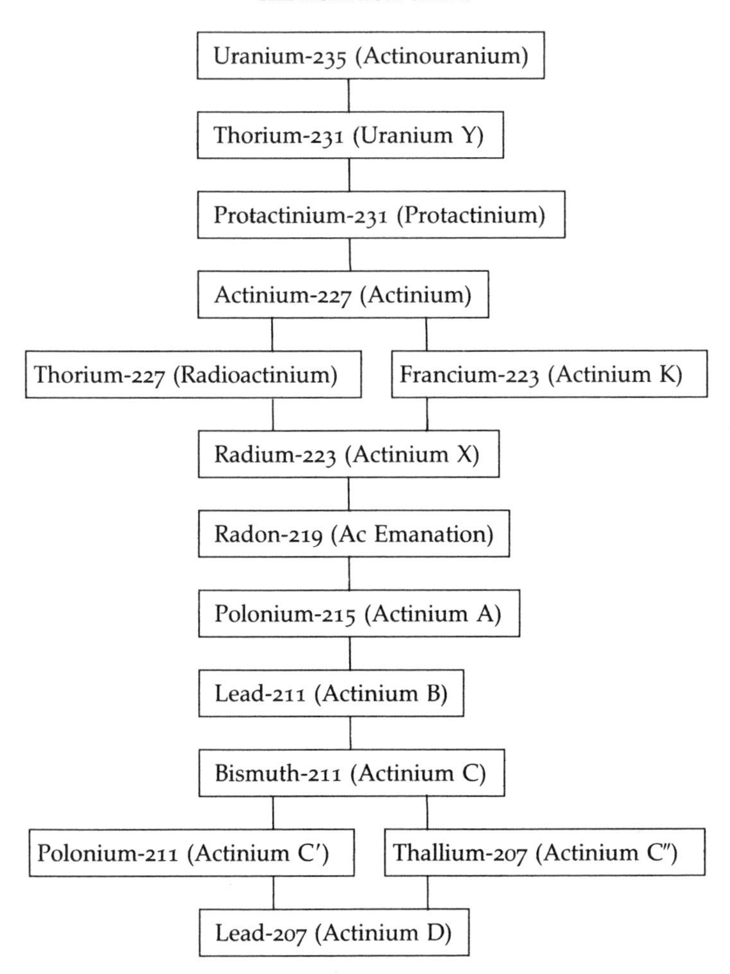

APPENDIX 2
THE CAVENDISH SONG

This version of the song, handwritten by Rutherford, was found among Brooks's memorabilia.

"An α Ray"
(*Air "A Jovial Monk"*)

An α ray was I contented with my lot
From Radium C I was set free
And outwards I was shot
My speed I quickly reckoned
As I flew off through space.
Ten thousand miles a second
Is quite a fine old pace.

For an α ray
goes a good long way
In a short time t
As you easily see
though I don't know why
My speeds so high
Or why I have a charge 2e.

On my wild career, as swiftly as I flew
A rarified gas wouldn't let me pass

But I shoved my way right through,
I had some lively tussles
To make it ionize
But I met the small corpuscles
A buzzing round like flies.

For an α ray
Hasn't time to stay,
while a low down gas
of inferior class,
Scarce conducting sound,
goes dancing around,
and plays the goat when I want to pass.

An electroscope looked on as I made that gas conduct
Beneath the field the gas did yield
And the leaf was gently "bucked"
But I murmured "Botheration"
A word that's most obscene
And I made a scintillation
As I struck a zinc-blend screen.

For an α ray
makes a fine display
with fluorescence green
on a zinc blend screen
when the room's quite dark,
you see a spark
that marks the spot where I have been.

But now I'm settled down and I move about quite slow
For I alas! am helium gas
Since I got that dreadful blow
but though I'm feeling sickly
Still no one now denies
That I ran that race so quickly
That I've won a Nobel Prize.

For an α ray
Is a thing to pay

and a Nobel Prize
one cannot despise
and Rutherford
has greatly scored
as all the world now recognize.

A Robb

APPENDIX 3

ADDRESS BY HARRIET BROOKS TO THE McGILL ALUMNAE SOCIETY

Perhaps it was a feeling of delicacy that prompted the committee to confine my attention to Mme Curie this p.m. that I might not be forced to expose the poverty of women's contributions to physical science. Even pure mathematics with such names as Mme Kovalevsky and Miss Scott of Bryn Mawr can make a braver showing. The combination of the ability to think in mathematical formulas and to manipulate skillfully the whimsical instruments of a physical laboratory a comb. [combination] necessary to attain eminence in physics is apparently one seldom met with in women. I may seem to run counter to generally accepted views in this but it has been my experience that men are as a rule much more skillful in the manipulation of delicate and intricate work than women. It is no rare thing to see a man with hands twice the size of one's own and every appearance of being very dangerous to fragile furniture who will handle quartz fibres so fine that they can be seen only against a black cloth, with the same ease with which we would disentangle a skein of wool.

Mrs Mary Somerville is probably the only other woman whose work can be classed with that of Mme Curie. She enjoyed the distinction according to Laplace of being the only woman who understood his "Mechanique Celeste" and her "Connection of the Physical Sciences" tho' written over half a century ago can still be read with pleasure and profit. That she was not deficient in experimental skill either is witnessed by the fact that she observed the magnetising effect of light rays, an effect that even with the more refined instruments of today is still too small for accurate measurement.

But I must not blame entirely the difficulties besetting the higher reaches of mathematics and our lack of deftness of hand for our neglect of the physical sciences. Their want of human interest has I think had much to do with our indifference to them. But the close of the old and the beginning of the new century has been marked by the astonishingly rapid development of a subject which has added a new philosophical interest to the phenomena of heat, light and electricity for our knowledge of radio-activity is bridging over the gulf separating matter and electricity. Thanks to the explanation of its phenomena we know that an atom of matter consists of a combination more or less simple, of moving charges of electricity, each atom constituting a closed system more or less stable. We have at the same time been brought back to the old theory that all matter is composed of the same primordial substance, the different properties of different substances being due to the grouping and number of the constituent parts of the atom.

Let me sketch briefly the development of this subject of radio-activity which in the last fifteen yrs has assumed such proportions that a separate course of lectures is now devoted to it in most Universities. In this development Mme Curie has played an important and unforgettable part.

The sensational discovery of the X Rays set investigators in all countries to work to try to find if any natural substances existed which emitted radiation that would penetrate metal and other substances opaque to light. The 1st important observation was made by Henri Becquerel of Paris – who found that a salt of Ur [uranium] – a rare earth of high at. wt. [atomic weight] gave an impression on a photographic enveloped in blk. [black] paper thus excluding all light rays. The rays that affected the plate were found to pass thro' thin sheets of metal and other substances. In addition to this Becquerel showed that the rays, in common with X rays, possessed the property of discharging an electrified body exposed to their action, a property of the highest importance for it has afforded a means a million times more delicate than the chemical balance, for determining the presence of minute quantities of the radio-elements. Mme Curie who was at that time teaching Chemistry at the Ecole Normal de Sévres and whose husband was Prof. of Physics at the Municipal School of Paris, made a detailed examination by the electrical method of the great majority of known substances. With the exception of Th. [thorium] a well known element largely used in manufacture of gas mantles, and a few ores such as chalsolite, carnotite and various samples of pitchblende, none possessed a radiation or an activity as generally called of even 1/100th that of Ur [uranium]. A similar activity occurring in several minerals made it seem probable that it was due to some active impurity. This was borne out by the fact that artificially prepared chalsolite was quite inactive. A pitchblende from Joachimsthal was the most active of the minerals examined and obtaining several tons of it from the Austrian government Mme Curie and her husband began the long and laborious separation which yielded finally a few decigrammes of the chloride of a substance which they called radium and which was enormously active as compared with Uranium. It was important to determine in the first place if the substance were really a new element or only a modification of some other. Further examination by the Curies shewed that it gave a new and characteristic line in the spectrum and its atomic wt. [weight] was ascertained to be 225, corrected with purer material within the last year by Mme Curie to 226.2. Another radio-active substance was isolated by Madame Curie from pitchblende and called by her polonium after her native land Poland. It has since been shewn to be one of the forms of radium. Through the kindness of Mme Curie specimens of RaCl [radium chloride] though still so rare were available to other investigators and thru the endeavors to solve the riddle of the new element, to shew whether this activity were an enduring and integral quality and to explain its nature. The greater part of this work was done here in our own laboratory by Prof. Rutherford, the French investigators including the Curies were slow to grasp the meaning of the phenomena they observed so well. They had it is true grasped an important principle the radiation as a property of the atom of the substance but their general view of the processes has not been substantiated.

Numberless difficulties naturally beset the work of separation of such extremely minute quantities of an unknown substance not the least of wh. [which] was the fact that when a small quantity of Ra. [radium] was, all the solid objects in the neighborhood assumed for a time the property of discharging electrified bodies, the test for the radio-active substances. On being removed however from the neighborhood of the element this power gradually disappeared. Prof. Rutherford had noticed the same phenomenon in connection with Th. [thorium] and showed that a gas was being continuously evolved from the Th. [thorium] which coming into contact with surrounding objects deposited a temporarily radio-active substance on their surfaces. Further examination by the Curies shewed that Ra. [radium] emitted a similar gas and thus was prepared the way for the theory of successive changes to explain these curious phenomena which today enjoys universal acceptance.

This theory may be briefly explained as follows, let us take Th. [thorium] as an example, any given sample possesses a certain amount of activity i.e. it can dissipate a given charge of electricity in a certain time. Treat this with Am. [ammonia] most of the Th. [thorium] is precipitated but the activity remains in the sol^{n} [solution] & the precipitate is inactive. Examined a month later the Th. [thorium] had regained all its activity and the sol^{n} [solution] was inactive. That is the Th. [thorium] had regrown the active const. [constituent] a substance christened Th. X [thorium X] with chemical properties quite distinct from the parent Th. [thorium]. Th X [thorium X] in its turn was found to be responsible for the emanation gas out of which was evolved the radio-active substance manifesting itself on surrounding objects, called Th. A [thorium A]. This in turn became Th. B [thorium B] before subsiding into activity as some unknown & so far undetectable substance. Radium has a much longer life history, in fact it is itself a product of Ur. [uranium] but the comparatively slow rate at wh. [which] it changes allows us to regard it as permanent. One gr. [gram] would dwindle to only 1/2 gramme in 1000 years. Radium changes first into a gas called the emanation which in turn becomes the solid Rad A [radium A] - - - - [through] Rad F [radium F] each with distinct chemical properties. Rad F [radium F] has recently been identified with the pol^{m} [polonium] of Mme Curie. The final inactive product of radium is thought to be lead and the recent separation by Mme Curie of a considerable quantity of pol^{m} [polonium] renders it quite possible that this question may soon be settled.

The test for a radio-active body is as the name indicates, the presence of a radiation. The X-rays had been of a simple type a sort of ethereal pulse of great penetrating power. Mme Curie early established the fact that the radiations from radium were quite different in character, they were more easily absorbed and they were not homogeneous. The greater part of the radiation was found to consist of positively charged particles projected with V [velocity] $= 1.65 \times 10^{4}$ cms per ″ [second] called α rays. In some cases as in that of pol^{m} [polonium] this was the only radiation emitted, but if any exists this type is always present. In many cases in addition to the α rays, negatively charged particles β rays are emitted with still greater velocity and accompanying them gamma rays, almost identical in character with X rays. Graphic method of representation of difference bet^{n} [between] rays due to Curies, Ra [radium] in hollow of lead block issuing pencils subjected to mag^{c} [magnetic] field. α rays deflected to one side β rays to the opposite & gamma rays pass upward unaffected. Each of these types of radiation possess distinct properties in the determination of wh. [which] Mme Curie's work was of great value the α rays have been shewn to be particles of helium a gas of low at. wt. [atomic weight] and as we have seen they accompany each transformation, it seems probable that when for instance the complex atom of Ra [radium] becomes unstable an atom of helium is thrown off and the remaining rearrange themselves forming a substance with different properties & a different period of stability. All radio-active bodies are of high atomic wt [weight] and ∴ [therefore] probably of complicated structure. Question naturally arises whether the process of the degradation of matter is a universal property, whether all substance are being gradually resolved into simpler forms as is observed in the hotter stars where hydrogen & helium prevail. If so such a process must be accompanied by other manifestations than those exhibited by the radio-elements and it must be very slow.

But to return to radium there is still one discovery of great importance which we must lay to the credit of the Curies. They observed that the substance maintained itself steadily at 1.5°C above the surrounding atmosphere. This heating effect is due to the bombardment of the Ra. [radium] particles by the α rays as they are repelled & in this way 1 gr. [gram] of Ra [radium] emits per day as much energy as is req'd [required] to dissociate 1 gr. [gram] of water. This discovery has afforded a most valuable means of measuring the energy of the radiations.

It is naturally difficult to separate the work of Madame Curie from that of her husband during the years in which they worked together but I have always had the impression that she was the pioneer in planning and he the skillful experimenter an impression confirmed by the opinions expressed by their collaborator when I was in Paris after the death of Prof. Curie. He had been appointed Professor of Physics at the Sorbonne in 1900 & met a most untimely death in April 1906 when he was crushed under a heavy van when crossing the Place Dauphine. He was a man evidently of very rare gifts and the instruments of exact measurement bearing his name bear testimony to his skill. In conjunction with his wife and Bequerel he was awarded the Nobel prize and refused the Cross of the Legion of Honor.

At his death his wife was appointed his successor at the Sorbonne, an honour to a woman quite without precedent at that ancient University. The duties of the position, the growth of the research laboratory under her care and no doubt also the loss of her skillful collaborator have somewhat lessened her output of work in the last few years. Physically she is extremely frail and her two children naturally absorb some of her attention, so that the fresh enthusiasm and interest with which she was ever ready to grapple with the difficulties and problems of those working with her, even when she was a prey to anxieties that would unnerve most women, were ever a source of the greatest admiration and wonder to me.

NOTES

CHAPTER ONE

1 Marianne G. Ainley, "Women Scientists in Canada: The Need for Documentation," *Resources for Feminist Research* 15 (November 1986): 7–8. See also Marianne G. Ainley, ed., *Despite the Odds* (Montreal: Véhicule Press, 1989).

2 H.J. Mozans, *Woman in Science* (1913; reprint, Cambridge, Mass.: MIT Press, 1974), 202.

3 A. Rebière, *Les femmes dans la science*, 2d ed. (Paris: Librairie Nony, 1897), 54. This book is an overlooked gold mine of information on several hundred women scientists from the earliest times up to the end of the nineteenth century. See also Carolyn Iltis, "Madame du Châtelet's Metaphysics and Mechanics," *Studies in History and Philosophy of Science* 8 (1977): 29–48.

4 Sally G. Kohlstedt, "In from the Periphery: American Women in Science, 1830–1880," *Signs* 4 (1978): 81–96.

5 John G. Reid, "The Education of Women at Mount Allison, 1854–1914," *Acadiensis* 12 (Spring 1983): 3–33.

6 Robin S. Harris, *A History of Higher Education in Canada, 1663–1960* (Toronto: University of Toronto, 1976), 116.

7 Ramsay Cook and Wendy Mitchinson, eds., *The Proper Sphere: Woman's Place in Canadian Society* (Toronto: Oxford University Press, 1976), 120.

8 Ruth Hubbard, "Feminism in Academia: Its Problematic and Problems," in Anne M. Briscoe and Sheila M. Pfafflin, eds., *Expanding the Role of Women in the Sciences* (New York: New York Academy of Sciences, 1979), 251. See also Joan N. Burstyn, "Education and Sex: The Medical Case against Higher Education for Women in England, 1870–1900," *Proceedings of the American Philosophical Society* 117 (1973): 79–89; and Louise M. Newman, "The Evils of Education (1870–1900)," in Louise M. Newman, ed., *Men's Ideas/Women's Realities: Popular Science, 1870–1915* (New York: Pergamon Press, 1985), 54.

9 Roy MacLeod and Russell Moseley, "Fathers and Daughters: Reflections on Women, Science and Victorian Cambridge," *History of Education* 8 (1979): 321–33.

10 Lydia E. Becker, "On the Study of Science by Women," *Contemporary Review* 10 (1869): 386–404.

11 Sir Ernest Rutherford, born in New Zealand in 1871, moved to England after graduating from Canterbury College, Christchurch. In England, he started work on the study of radioactivity, and as a result of his outstanding research, he was offered a professorship at McGill University in 1898. It was at McGill that Brooks worked with him. Rutherford moved to Manchester University in England in 1907 and then back to Cambridge University in 1919. Rutherford is best known for his work on the structure of the atom for which he was awarded the Nobel Prize for chemistry in 1908. He died in 1937.

12 Margaret W. Rossiter, "Sexual Segregation in the Sciences: Some Data and a Model," *Signs* 4 (1988): 146–51.

13 Kathleen Lonsdale, "Women in Science: Reminiscences and Reflections," *Impact of Science on Society* 20 (January–March 1970): 45–59. The crucial role of Bragg in encouraging women into crystallography is discussed in Maureen M. Julian, "Women in Crystallography," in G. Kass-Simon and Patricia Farnes, eds., *Women of Science* (Bloomington: Indiana University Press, 1990), 335–83.

14 M.M. Julian, "Dame Kathleen Lonsdale," *Physics Teacher* 19 (March 1981): 159–65.

15 M.M. Julian, "Dorothy Crowfoot Hodgkin: Nobel Laureate," *Journal of Chemical Education* 59 (1982): 124–5.

16 Rosalind Franklin, who played a major role in the discovery of the structure of DNA, moved into the field of crystallography under the

tutelage of one of Braggs's former students. See M.M. Julian, "Rosalind Franklin: From Coal to DNA to Plant Viruses," *Journal of Chemical Education* 60 (1983): 660–2.

17 Bertram Boltwood, for example, did not want Ellen Gleditsch to work with him, initially considering her a "husband-hunter." See Lawrence Badash, ed., *Rutherford and Boltwood: Letters on Radioactivity* (New Haven, Conn.: Yale University Press, 1969), 285.

18 Marelene F. Rayner-Canham and Geoffrey W. Rayner-Canham, "Pioneer Women in Nuclear Science," *American Journal of Physics* 58 (1990): 1036–43.

19 Ernest Rutherford, *Radioactivity*, 2d ed. (Cambridge: Cambridge University Press, 1905); Ernest Rutherford, *Radioactive Substances and Their Transformations* (Cambridge: Cambridge University Press, 1913).

20 Ernest Rutherford, "The Succession of Changes in Radioactive Bodies," *Philosophical Transactions of the Royal Society*, series A, 204 (1904): 169–219.

21 Ellen Gleditsch to Ernest Rutherford, 1 November 1915, Rutherford Collection of Correspondence (RCC).

22 Born Marya Sklodovska in Poland in 1867, Curie first learned about science from books. She enrolled at the Sorbonne in 1891 and obtained master's degrees in physics (1893) and mathematics (1894). Her first research work was with G. Lippman. For her doctorate she chose the field of radioactivity, a phenomenon just discovered by A. Becquerel. Through research she undertook with her spouse, Pierre Curie, she discovered two new elements, polonium and radium. After her husband's accidental death in 1906, M. Curie was granted P. Curie's position at the Sorbonne. She continued to work on the chemistry of radium, but as well, she devoted considerable time to the medical use of X-rays. Curie was awarded the Nobel Prize for physics in 1903 and for chemistry in 1911. She died in France in 1934.

23 Bertram Boltwood was an American scientist born at Amherst, Massachusetts, in 1870. After graduating from Yale University, he undertook postgraduate studies at Munich, Leipzig, and Manchester, the last under Ernest Rutherford. He then obtained a permanent post at Yale University. He is best known for devising the use of radioactivity in the dating of geological samples. Boltwood died in 1927.

24 Marie Curie to Ernest Rutherford, RCC, undated, cited in Robert Reid, *Marie Curie* (New York: Saturday Review Press, 1974).

25 Helena M. Pycior, "Marie Curie's 'Anti-natural Path,'" in Pnina G. Abir-Am and Dorinda Outram, eds., *Uneasy Careers and Intimate Lives* (New Brunswick: Rutgers University Press, 1987), 191. This quote has been used elsewhere to indicate Rutherford's attitude to women; see William Booth, "Oh, I Thought You Were a Man," *Science* 243 (1989): 475.

26 Lise Meitner, born in Vienna, Austria, in 1878, studied physics at the University of Vienna. After receiving her doctorate in physics, she moved to Berlin to work with Max Planck. It was there that she met Otto Hahn, and over the following decades they worked together on a number of problems related to radioactivity. In 1918 she was appointed head of physics at the prestigious Kaiser-Wilhelm Institute. Meitner was the co-discoverer of the element protactinium, but she made her most famous discovery with her nephew, O. Frisch – the phenomenon of nuclear fission. This work was done in Sweden, as she had had to flee Germany, being of Jewish descent. She retired to Cambridge, England, where she died in 1968. Many consider that the Nobel Prize for chemistry in 1944 awarded to Otto Hahn should have been given jointly to her.

27 Katherine Haramundanis, ed., *Cecilia Payne-Gaposchkin* (Cambridge: Cambridge University Press, 1984), 118.

28 These women scientists were Winifred Moller Beilby, Lucie Blanquies, Ada Florence Hitchins, Stephanie Horovitz, Elizabeth Rebecca Laird, Lise Meitner, Ruth Pirret, Eva Ramstedt, Jessie Mable Wilkins Slater, and Edith Gertrude Willcock. See note 18.

29 Joyce Antler, '"After College, What?': New Graduates and the Family Claim," *American Quarterly* 32 (Fall 1980): 409–34.

30 Margaret W. Rossiter, "'Women's Work' in Science, 1880–1910," *ISIS* 71 (1980): 381–98; also in Margaret W. Rossiter, *Women Scientists in America* (Baltimore: Johns Hopkins University Press, 1982), 52.

31 Deborah J. Warner, "Women Astronomers," *Natural History* 88 (May 1979): 12–26.

32 Marion Talbot, "Eminence of Women in Science," *Science* 32 (1910): 866.

33 Penina M. Glazer and Miriam Slater, *Unequal Colleagues: The Entrance of Women into the Professions, 1890–1940* (New Brunswick: Rutgers University Press, 1986), 19.

34 Daniel J. Kevles, *The Physicists: The History of a Scientific Community in Modern America* (Cambridge, Mass.: Harvard University Press, 1987), 207. Most of the physics faculty at Princeton would not even allow Dewey to take up a research fellowship there. The dean had to intervene on her behalf.
35 Elaine Kendall, *Peculiar Institutions* (New York: G.P. Putnam, 1976), 127.
36 Alison Prentice et al., *Canadian Women: A History* (Toronto: Harcourt Brace Jovanovich, 1988), 159.
37 Jessie Bernard, *Academic Women* (University Park: Pennsylvania State University Press, 1964), 207.
38 Sir Joseph John Thomson was born in Manchester, England, in 1856. He started his studies at Owen's College (later the University of Manchester) at the age of fourteen and then continued studying mathematics at Cambridge University. After graduating, he obtained a post at Cambridge, where he remained until his death in 1940. Thomson discovered the electron and determined its charge to mass ratio. His group of students at the Cavendish Laboratory, Cambridge, included Ernest Rutherford and, later, Harriet Brooks. He received the Nobel Prize for physics in 1906.
39 Evelyn Fox Keller, "Women in Science," *Harvard Magazine* 77 (1974): 14–19.
40 Burstyn notes that of the 218 married women who graduated from Smith College during the period 1879–99, only 10 continued to work after marriage. See Joan N. Burstyn, "American Society during the Eighteen Nineties: 'The Woman Question,'" *Studies in History and Society* 4 (1972): 34–40.
41 Jonathan R. Cole, *Fair Science* (New York: Free Press, 1979), 249.

CHAPTER TWO

1 Nick Mika and Helma Mika, *Places in Ontario*, part 1 (Belleville, Ont.: Mika, 1977), 704–5.
2 We are indebted to Cicely Grinling for providing us with extracts from the reminiscences of Elizabeth (Brooks) Eve, for listings from the Worden and Brooks family Bibles, and for her own comments on the family history.
3 Robin Stanes, *A History of Devon* (Chichester, UK: Phillimore, 1986), 111.

4 Mika and Mika, *Places in Ontario*, part 1, 1333. The village was originally called Mumby's Mills, after the mill owner. A dispute that Mumby had with a Scotsman was said to have been like the Battle of Bannockburn. The settlement adopted this name officially in the 1860s.
5 We thank Dosie Pitcher for this suggestion.
6 Nick Mika and Helma Mika, *Places in Ontario*, part 2 (Belleville, Ont.: Mika, 1981), 46–47. We know the population in 1888 was 2,000 inhabitants, as this was the number required for incorporation. So that this headcount could be reached, it was arranged for one of the Grand Trunk Railway passenger trains to remain at the station for one hour – long enough to register the passengers and crew as Forest citizens.
7 Nick Mika and Helma Mika, *Places in Ontario*, part 3 (Belleville, Ont.: Mika, 1983), 369–71.
8 Mika and Mika, *Places in Ontario*, part 3, 369–71.
9 Howard W. James, head of guidance, Seaforth Collegiate Institute, is thanked for his efforts in tracking down the early school records.
10 Ellen Hayes, "Women and Scientific Research," *Science* 32 (1910): 864–6. It is interesting and somewhat depressing that the same views are being echoed eighty years later. See Barbara Lloyd, "Rules of the Gender Game," *New Scientist* 124 (2 December 1989): 60–4.
11 Reminiscences of Elizabeth (Brooks) Eve.
12 Arthur Stewart Eve was born in Sisoe, England, in 1862. After teaching at Marlborough College, England, he transferred to McGill University in 1903, where he performed research with Rutherford. Eve remained at McGill for the rest of his active life and was promoted to the position of director of the physics department. Apart from his work on radioactivity, he is known for his research in geophysics. Eve died in 1948.
13 Henry J. Morgan, ed., *The Canadian Men and Women of the Time* (Toronto: William Briggs, 1912), 455.

CHAPTER THREE

1 M. Gillett, *We Walked Very Warily* (Montreal: Eden Press Women's Publications, 1981). This book gives more extensive coverage of the entry of women into McGill University.

2 Jill Conway, "Stereotypes of Femininity in a Theory of Sexual Evolution," in Martha Vicinus, ed., *Suffer and Be Still: Women in the Victorian Age* (Bloomington: Indiana University Press, 1972), 141.
3 Barbara M. Solomon, *In the Company of Educated Women* (New Haven, Conn.: Yale University Press, 1985), 63.
4 Solomon, *In the Company of Educated Women*, 63.
5 Judith Fingard, "College, Career, and Community: Dalhousie Coeds, 1881–1921," in Paul Axelrod and John G. Reid, eds., *Youth, University and Canadian Society* (Montreal/Kingston: McGill-Queen's University Press, 1989), 26.
6 M. Carey Thomas, "Present Tendencies in Women's College and University Education," *Educational Review* 35 (1908): 64–85; reproduced in J. Stacey et al., eds., *And Jill Came Tumbling After – Sexism in American Education* (New York: Laurel, 1974), 276.
7 Martha Vicinus, *Independent Women: Work and Community for Single Women 1850–1920* (Chicago: University of Chicago Press, 1985), 138.
8 Gillett, *We Walked Very Warily*.
9 C.M. Derick, "In the 80's," in *Old McGill 1927*, 200.
10 Elizabeth A. Irwin, "Women at McGill," *McGill News* 1 (December 1919): 41.
11 John Satterly, "Reminiscences in Physics from 1894 Onward," *American Journal of Physics* 25 (1957): 288–300.
12 We thank Phebe Chartrand, archivist, McGill University, for collecting and collating the information on the courses that Brooks took as an undergraduate.
13 *Old McGill 1898*, 51.
14 Gillett, *We Walked Very Warily*, 341.
15 D. Suzanne Cross, "The Neglected Majority: The Changing Role of Women in 19th Century Montreal," *Histoire Sociale/Social History* 6 (12 November 1973): 202–23. It is unclear whether Brooks obtained exemption from this rule or whether her later tutorship at the Royal Victoria College was taken as satisfying this obligation.
16 Cyrus MacMillan, *McGill and Its Story 1821–1921* (London: Oxford University Press, 1921), 238.
17 Lewis Pyenson, "The Incomplete Transmission of a European Image: Physics at Greater Buenos Aires and Montreal, 1890–1920," *Proceedings of the American Philosophical Society* 122 (1978): 92–114.
18 Gloria B. Lubkin, "Women in Physics," *Physics Today* 24 (April 1971): 23–7.

19 A. Stewart Eve, *Rutherford* (Cambridge: Cambridge University Press, 1939).
20 Norman Feather, *Lord Rutherford* (London: Priory Press, 1940), 63.
21 P.L. Kapitza, "Recollections of Lord Rutherford," *Royal Society (Great Britain), Proceedings*, series A, A294 (1966): 123–37.
22 Harriet Brooks, "Damping of Electrical Oscillations," *Royal Society of Canada (Transactions)*, section 3 (1899), 13–15.
23 Harriet Brooks, "Damping of the Oscillations in the Discharge of a Leyden Jar" (M.A. thesis, McGill University, 1901). The first women to obtain master's degrees at McGill were Elizabeth Binmore (botany, 1894) and Euphemia McLeod (philosophy, 1894), while Annie L. Macleod (chemistry, 1910) was the first woman to obtain a doctoral degree. See Gillett, *We Walked Very Warily*, 419.
24 McGill University Archives, *Minute Book of the Faculty of Arts*. We thank Phebe Chartrand, archivist, McGill University, for supplying these citations.
25 A. Norman Shaw, "Recollections of Robert William Boyle 1883–1955: A Distinguished Canadian Physicist," *Physics in Canada* 10 (1955): 21–8.
26 Eve, *Rutherford*, 69. Rutherford wrote his wife-to-be, Mary Newton: "I am very busy with research work just now, I have three men going at research and one lady, I have to look after them very closely."
27 *Old McGill 1899–1900*, 141.
28 Gillett, *We Walked Very Warily*.
29 M.V. Roscoe, *The Royal Victoria College 1899–1962 – A Report to the Principal* (Montreal: McGill University, 1964), 6. The hiring of Brooks is noted in the board of governors minutes for 23 September 1899: "... and Miss Harriet Brooks as non-Resident Tutor in Mathematics in the Royal Victoria College at a salary of $300 for the session."
30 *Gazette* (Montreal), 2 November 1900.
31 H.D. Oakley, *My Adventures in Education* (London: Williams and Norgate, 1939), 85.
32 G.E.M. Jauncey, "The Early Years of Radioactivity," *American Journal of Physics* 14 (1946): 226–41.
33 S.B. Sinclair, "J.J. Thomson and Radioactivity: Part I," *Ambix* 35 (1988): 91–104.
34 E. Rutherford, "A Radioactive Substance Emitted from Thorium Compounds," *Philosophical Magazine*, series 5, 49 (1900): 1–14.

35 We would now identify the gas as being the radioactive gaseous element radon. Due to the inherent errors of the method, the value that Brooks obtained for the molecular weight was very low. However, this does not invalidate the significance of her work.

36 Ernest Rutherford and Harriet T. Brooks, "The New Gas from Radium," *Royal Society of Canada (Transactions)*, section 3 (1901); 21–5; also published in *Chemical News*, 25 April 1902, 196–7. Her first task may have been to help Rutherford determine the effect of temperature change on the production of "emanation" from thorium, as Rutherford thanks Brooks for her help in his paper "Einfluss der Temperatur auf die 'Emanationen' radioaktiver Substanzen," *Physikalische Zeitschrift* 2 (1901): 429–31.

37 E. Rutherford, "Emanations from Radio-active Substances," *Nature* 64 (1901): 157–8.

38 We thank Dr Montague Cohen, McGill University, for pointing out to us the importance of Brooks's discovery.

39 Marjorie Malley, "The Discovery of Atomic Transmutation: Scientific Styles and Philosophies in France and Britain," *ISIS* 70 (1979): 213–23.

40 D. Wilson, *Rutherford* (Cambridge, Mass.: MIT Press, 1983), 225.

41 Wilson, *Rutherford*, 46.

42 H.R. Robinson, "Rutherford: Life and Work to the Year 1919, with Personal Reminiscences of the Manchester Period," *Physical Society (Proceedings)* 55 (1943): 161; reprinted in *Rutherford by Those Who Knew Him* (London: Physical Society, 1954), 16.

43 Wilson, *Rutherford*, 264.

44 *Times* (London), 8 December 1920, 8.

45 Ernest Rutherford and Harriet T. Brooks, "Comparison of the Radiations from Radioactive Substances," *Philosophical Magazine*, series 6, 4 (July 1902): 1–23.

46 Ernest Rutherford, *Radioactive Substances and Their Transformations* (Cambridge: Cambridge University Press, 1913), 487.

47 F.R. Terroux, "The Rutherford Collection of Apparatus at McGill University," *Royal Society of Canada (Proceedings and Transactions)*, series 3, 32 (1938): 9–16.

48 One of the authors (G. R-C) thanks Montague Cohen, curator of the Rutherford Collection, for a personal tour of the display.

49 W. Peterson to H. Brooks, 22 March 1901, McGill University archives. We thank Phebe Chartrand, Archivist, McGill University, for copies of the correspondence of Peterson and Brooks.

CHAPTER FOUR

1 H.L. Horowitz, *Alma Mater* (New York: Alfred A. Knopf, 1984), 115.

2 Roberta Wein, "Women's Colleges and Domesticity, 1875–1918," *History of Education Quarterly* 14 (1974): 31–47.

3 Sonya (Sophia) Kovalevsky (1850–91) was an outstanding Russian mathematician. Her struggles to become a mathematician are fascinating. So that they could go abroad and study, it was the custom in her time for intelligent women to contract a marriage of convenience with a man who was also planning to go to a foreign university. Kovalevsky's particularly well-known works were on partial differential equations, Abelian integrals, and the rings of the planet Saturn. A detailed discussion of her life is given in Ann Hibner Koblitz, *A Convergence of Lives: Sofia Kovalevskaia, Scientist, Writer, Revolutionary* (Cambridge, Mass.: Birkhäuser, 1983). Koblitz argues that Kovalevsky was the most outstanding woman scientist prior to the twentieth century: Her personal life is the focus of Don H. Kennedy, *Little Sparrow: A Portrait of Sophia Kovalevsky* (Athens, Ohio: Ohio University Press, 1983), while her work in mathematics is discussed in Roger Cooke, *The Mathematics of Sonya Kovalevskaya* (New York: Springer-Verlag, 1984).

4 M. Carey Thomas, "Present Tendencies in Women's College and University Education," *Educational Review* 35 (1908): 64–85.

5 C. Meigs, *What Makes a College?* (New York: Macmillan, 1956), 32.

6 Helen T. Flexner, *Bryn Mawr – A Characterisation* (Bryn Mawr, 1905), 12.

7 We thank Lucy Fisher West, college archivist, Bryn Mawr College, for obtaining Brooks's transcript and Caroline Rittenhouse, college archivist, Bryn Mawr, for obtaining the physics and mathematics sections of the Bryn Mawr College Graduate Program, 1901–1902.

8 Derek J. Price, "The Cavendish Laboratory Archives," *Royal Society, London (Notes and Records)* 10 (1953): 139–47; and Lawrence Badash, *Rutherford Correspondence Catalog* (New York: Center for History of Physics, 1974).

9 Copies of the relevant letters were obtained from the Rutherford Collection of Correspondence (RCC). We thank A.E.B. Owen, keeper of manuscripts, Cambridge University Library, for assistance.

10 Edith Finch, *Carey Thomas of Bryn Mawr* (New York: Harper & Brothers, 1947), 166.
11 H. Brooks to E. Rutherford, 8 December 1901, RCC.
12 Sheila E. Widnall, "AAAS Presidential Lecture: Voices from the Pipeline," *Science* 241 (1988): 1740–5.
13 The paper to which she refers is the major work on the comparison of radiations from radioactive substances: Ernest Rutherford and Harriet T. Brooks, "Comparison of the Radiations from Radioactive Substances," *Philosophical Magazine*, series 6, 4 (July 1902): 1–23.
14 H. Brooks to E. Rutherford, 18 March 1902, RCC.
15 Flexner, *Bryn Mawr – A Characterisation*, 12.
16 Martha Vicinus, "One Life to Stand beside Me: Emotional Conflicts in First Generation College Women in England," *Feminist Studies* 8 (Fall 1982): 603–28.
17 Hugh MacLennan, *McGill* (London: Allen and Unwin, 1960), 81.
18 H. Brooks to E. Rutherford, 2 April 1902, RCC.
19 W. Peterson to H. Brooks, 8 April 1902, McGill Archives. We thank Phebe Chartrand, archivist, McGill University, for copies of this and later correspondence of Principal Peterson.
20 A. Stewart Eve, *Rutherford* (Cambridge: Cambridge University Press, 1939), 81.
21 Bryn Mawr College, Graduate Program, 1901–1902.
22 H. Brooks to E. Rutherford, 17 April 1902, RCC.
23 We thank A.E.B. Owen, keeper of manuscripts, Cambridge University Library, for a fruitless search for this communication.
24 J.J. Thomson to E. Rutherford, 13 May 1902, RCC.
25 H. Brooks to E. Rutherford, 27 May 1902, RCC.
26 Daniel J. Kevles, *The Physicists: The History of a Scientific Community in Modern America*, 2d ed. (Cambridge, Mass.: Harvard University Press, 1987), 76.
27 The Mrs Sidgwick to whom Brooks refers was physics researcher Eleanor Balfour, spouse of Cambridge moral philosopher Henry Sidgwick. Balfour collaborated with Lord Rayleigh (the Elder) in much of his research into electricity. Later she became principal of Newnham College, and it was in this capacity that Brooks wrote to her. For a complete biography, see Ethel Sidgwick, *Mrs Henry Sidgwick* (London: Sidgwick and Jackson, 1938).
28 H. Brooks to E. Rutherford, 27 May 1902, RCC.

CHAPTER FIVE

1 J.G. Crowther, *The Cavendish Laboratory 1874–1974* (New York: Science History Publications, 1974), 31.
2 Lawrence Bragg, "Famous Experimental Apparatus in the Cavendish Laboratory, Cambridge," *Nature* 166 (1950): 7–9.
3 Mary A. Hamilton, *Newnham: An Informal Biography* (London: Faber and Faber, 1936).
4 M.A.R. Tuker, *Cambridge* (London: Adam and Charles Black, 1907).
5 Mary Phillips, librarian, Newnham College, is thanked for searching the minute book and for providing this information.
6 This information was obtained from records at Bryn Mawr. As holder of a European fellowship, Brooks had to present a report to the president and faculty of Bryn Mawr College upon completion of her year abroad. We thank Caroline Rittenhouse, archivist, Bryn Mawr College, for a copy of this document.
7 *A History of the Cavendish Laboratory, 1871–1910* (London: Longmans, Green, and Co., 1910), 225.
8 Sheila E. Widnall, "AAAS Presidential Lecture: Voices from the Pipeline," *Science* 241 (1988): 1740–5.
9 George P. Thomson, *J.J. Thomson and the Cavendish Laboratory in His Day* (New York: Doubleday, 1965), 91.
10 The borrowing of ideas from Germany was quite prevalent during this period. See George Haines, *Essays on German Influence upon English Education and Science 1850–1919* (Hamden: Connecticut College/Archon Books, 1969).
11 Thomson, *J.J. Thomson and the Cavendish Laboratory*, 91.
12 Georgina Ferry and Jane Moore, "True Confessions of Women in Science," *New Scientist* 95 (1 July 1982): 27–30.
13 Thomson, *J.J. Thomson and the Cavendish Laboratory*, 77.
14 Lord Rayleigh, *The Life of Sir J.J. Thomson* (Cambridge: Cambridge University Press, 1943), 40.
15 Rayleigh, *The Life of Sir J.J. Thomson*, 53.
16 E. Rutherford to J.J. Thomson, 26 December 1902, Rutherford Collection of Correspondence (RCC): "I hope Miss Brooks and McClung are doing well." Robert K. McClung, Rutherford's other pioneer student, had also obtained a scholarship to the Cavendish. We thank A.E.B. Owen, keeper of manuscripts, Cambridge University Library, for assistance in obtaining copies of the correspondence.

17 H. Brooks to E. Rutherford, 13 February 1903, RCC.
18 S.B. Sinclair, "J.J. Thomson and Radioactivity: Part I," *Ambix* 35 (July 1988): 91–104.
19 S.B. Sinclair, "J.J. Thomson and Radioactivity: Part II," *Ambix* 35 (November 1988): 113–26.
20 H. Brooks to E. Rutherford, 13 February 1903, RCC.
21 Rayleigh, *The Life of Sir J.J. Thomson*, 51.
22 Rayleigh, *The Life of Sir J.J. Thomson*, 28.
23 Rayleigh, *The Life of Sir J.J. Thomson*, 29.
24 J.J Thomson to E. Rutherford, 14 April 1903, RCC.
25 H. Brooks to E. Rutherford, undated, RCC.
26 Brooks's address in Berlin was Potsdamer Straβe 15. It may have been a coincidence, but a former student of J.J. Thomson's, Elsa Neumann, had lived at #10 in the same street prior to her death in 1902.
27 Slater was a student at Newnham College, Cambridge University, from 1899 to 1903 and at Bathurst College, Cambridge, from 1903 to 1905. Slater obtained a BSC degree (1902) and a DSC degree (1906) from the University of London – presumably because Cambridge would not formally grant degrees to women until 1948. For additional biographical information on Slater, see Marelene F. Rayner-Canham and Geoffrey W. Rayner-Canham, "Pioneer Women in Nuclear Science," *American Journal of Physics* 58 (1990): 1036–43.
28 Willcock was born 7 January 1879 at Albrighton, Salop, England. She was a student at Newnham College, Cambridge, from 1900 to 1904. For additional biographical information on Willcock, see Rayner-Canham and Rayner-Canham, "Pioneer Women in Nuclear Science," 1036–43.
29 J.M.W. Slater, "On the Excited Activity of Thorium," *Philosophical Magazine* 9 (1905): 628–44; J.M.W. Slater, "On the Emission of Negative Electricity by Radium and Thorium Emanations," *Philosophical Magazine* 10 (1905): 460–6.
30 The handwriting of this song was found to be the same as that in a number of letters from Rutherford to Schuster.
31 *Post-Prandial Proceedings of the Cavendish Society*, 6th ed. (Cambridge: Bowes and Bowes, 1926), 22.
32 John Satterly, "The Postprandial Proceedings of the Cavendish Society," *American Physics Teacher* 7 (1939): 179–85, 244–8.
33 Rutherford wrote to Bertram Boltwood at Yale University about the occasion and mentioned the song. See Lawrence Badash, ed., *Ruth-*

erford and Boltwood: Letters on Radioactivity (New Haven, Conn.: Yale University Press, 1969), 205. The minor differences between Brooks's and Boltwood's versions may reflect either a degree of illegibility in the version sent to Boltwood or errors made by Rutherford when he copied the verses.

34 W. Peterson to E. Rutherford, 18 March 1903, McGill University archives. We thank Phebe Chartrand, Archivist, McGill University, for copies of the correspondence of Peterson to Rutherford and to Brooks.

35 *Minutes of the Board of Governors Meeting,* 17 April 1903, McGill University Archives.

36 This appears to be quite a reasonable salary for the time. As Pyenson notes, in 1905 the highest-paid physics lecturers earned $750, while Rutherford himself received $4,000. See Lewis Pyenson, "The Incomplete Transmission of a European Image: Physics at Greater Buenos Aires and Montreal, 1890–1920," *American Philosophical Society (Proceedings)* 122 (1978): 92–114.

37 W. Peterson to H. Brooks, 5 May 1903, RCC.

CHAPTER SIX

1 Ernest Rutherford, "Early Days in Radio-Activity," *Journal of the Franklin Institute* 198 (September 1924): 281–90.

2 Ernest Rutherford and Harriet T. Brooks, "Comparison of the Radiations from Radioactive Substances," *Philosophical Magazine,* series 6, 4 (July 1902): 1–23.

3 Otto Hahn, born in Frankfurt am Main, Germany, in 1879, studied at Marburg University and then moved to England to work with Sir William Ramsay at University College, London. It was there that he first started research in radioactivity. From there, he spent a year at McGill with Rutherford and then returned to Germany to work at the University of Berlin. Hahn was a chemist and experimentalist. It was his fruitful work with the physicist and theoretician Lise Meitner, particularly on the products of nuclear fission, that led to his Nobel Prize for chemistry in 1944. Hahn died in Germany in 1968.

4 Otto Hahn, "Some Reminiscences of Professor Ernest Rutherford during His Time at McGill University, Montreal, Part 2," included in J. Chadwick, ed., *The Collected Papers of Lord Rutherford of Nelson* (London: Allen and Unwin, 1962), 164.

5 Harriet Brooks, "A Volatile Product from Radium," *Nature* 70 (21 July 1904): 270.
6 Frederick Soddy, *The Interpretation of Radium*, 4th ed. (New York: G.B. Putnam, 1920), 138.
7 A. Stewart Eve, *Rutherford* (Cambridge: Cambridge University Press, 1939), 129.
8 Ernest Rutherford, "The Succession of Changes in Radioactive Bodies," *Philosophical Transactions of the Royal Society*, series A, 204 (1904): 169–219.
9 Thaddeus J. Trenn, "Rutherford and Recoil Atoms: The Metamorphosis and Success of a Once Stillborn Theory," *Historical Studies in Physical Sciences* 6 (1975): 513–47.
10 Trenn, "Rutherford and Recoil Atoms," 513–47.
11 The only book to acknowledge the credit due to Brooks is M E. Weeks and H.M. Leicester, *Discovery of the Elements* (Easton, Pa.: Journal of Chemical Education, 1968), 786. Interestingly, Brooks is not mentioned in the text of the classic work on the history of chemistry by Ihde, but in a table ("Discovery of Radioactive Isotopes") in this work, Rutherford and Brooks are listed as the discoverers of radium A (polonium-218) and radium B (lead-214) during 1902–1904, and of radium C (bismuth-214), radium D (lead-210), radium E (bismuth-210), and radium F (polonium-210) during 1904–1905. See Aaron J. Ihde, *The Development of Modern Chemistry* (New York: Harper & Row, 1964), 750.
12 Otto Hahn, *Otto Hahn: A Scientific Biography* (New York: Charles Scribner, 1966), 62.
13 E. Rutherford to O. Hahn, 22 December 1908, Rutherford Collection of Correspondence (RCC).
14 Ernest Rutherford, *Radioactivity*, 2d ed. (Cambridge: Cambridge University Press, 1905), 392. "Since [radium] A breaks up with an expulsion of an α particle, some of the residual atoms constituting radium B may acquire sufficient velocity to escape into the gas, and are then transferred by diffusion to the walls of the vessel."
15 Interestingly, an appendix in Hahn's autobiography (*Otto Hahn: A Scientific Biography*) contains brief biographical notes on important individuals. In the note on Brooks (268), this comment is made: "She may have been the first researcher to have observed the phenomenon of radioactive recoil."

16 Harriet Brooks, "The Decay of the Excited Radioactivity from Thorium, Radium, and Actinium," *Philosophical Magazine* series 6, 8 (September 1904): 373–84.
17 Rutherford, "The Succession of Changes in Radioactive Bodies," 169–219.
18 One of the few to recognize her contribution was Mary E. Weeks in her "The Discovery of the Elements. XIX. The Radioactive Elements," *Journal of Chemical Education* 10 (1933): 79–90.
19 Brooks, "The Decay of the Excited Radioactivity from Thorium, Radium, and Actinium," 373–84.
20 A. Stewart Eve, "Some Scientific Centres. VIII. The Macdonald Physics Building, McGill University, Montreal," *Nature* 74 (1906): 272–5.
21 *Minutes of the McGill Physical Society*, 1897–1904, McGill University Archives.
22 Mary Violette Dover was a chemist who received her BA in 1898, the same year as Brooks. She was a demonstrator in chemistry at McGill until 1905. Dover obtained a PhD in 1908 from Breslau and became an instructor at Mt Holyoke College. Subsequently, she became an assistant professor of chemistry at the University of Missouri, Columbia, Missouri. Bella Marcuse obtained her BA from McGill in 1900 and MSc in 1903. She must have discontinued her work, as the only subsequent information about her refers to her marriage to Douglas McIntosh. McIntosh was a chemistry demonstrator at McGill at the time. He later became professor of chemistry at the University of British Columbia. We thank Phebe Chartrand, archivist, McGill University, for the information on Dover and Marcuse.
23 Fanny Cook Gates had degrees from Northwestern University and then held fellowships in physics and mathematics at Bryn Mawr College, McGill, Cambridge, Göttingen, Zurich, and Chicago. For additional biographical details, see Marelene F. Rayner-Canham and Geoffrey W. Rayner-Canham, "Some Pioneer Women in Nuclear Science," *American Journal of Physics* 58 (1990): 1036–43.
24 *Minutes of the McGill Physical Society*, 1897–1904.
25 W. Peterson to H. Brooks, 12 March 1904, McGill University Archives. We thank Phebe Chartrand, archivist, for delving into the correspondence of Principal Peterson to find this letter.

CHAPTER SEVEN

1 H.L. Horowitz, *Alma Mater* (New York: Alfred A. Knopf, 1984), 248.
2 Herbert M. Richards, "The Curriculum and the Equipment of Barnard College," *Columbia University Quarterly* 12 (March 1910): 172–80.
3 We thank Patricia K. Ballou, archivist, Barnard College, for copies of Brooks's letters of appointment and for extracts from the Barnard College Announcement listing Brooks's course assignments.
4 Harold W. Webb, "Bergen Davis," in *Biographical Memoirs*, vol. 34 (New York: National Academy of Sciences, Columbia University Press, 1960), 65.
5 We thank Lucinda Manning, archivist, Barnard College, for copies of these letters, which are filed as 41 Dean's Office Corresp. (1906–1908) Dept. 06–08.
6 M.C. White, *Barnard College* (New York: Columbia University Press, 1954), 65.
7 H. Brooks to L. Gill, 10 July 1906, Barnard College Archives (BCA).
8 White, *Barnard College*, 44.
9 A.D. Miller and S. Myers, *Barnard College* (New York: Columbia University Press, 1939), 63.
10 L. Gill to H. Brooks, 12 July 1906, BCA.
11 H. Brooks to L. Gill, 18 July 1906, BCA.
12 L. Gill to H. Brooks, 23 July 1906, BCA.
13 Virginia C. Gildersleeve, *Many a Good Crusade* (New York: Macmillan, 1954), 52.
14 M. Maltby to L. Gill, 24 July 1906, BCA.
15 M.B. Ogilvie, *Women in Science* (Cambridge, Mass.: MIT Press, 1986), 124; Patricia J. Siegel and Kay T. Finley, *Women in the Scientific Search: An American Bio-bibliography 1724–1979* (Metuchen, NJ: Scarecrow Press, 1985), 286.
16 L. Gill to M. Maltby, 30 July 1906, BCA.
17 Gildersleeve, *Many a Good Crusade*, 105.
18 Grace Langford became instructor in physics at Barnard in 1908 and stayed until at least 1920, according to *American Men in Science*, 1921 edition. Margaret W. Rossiter notes that Langford did not complete her PhD begun at Columbia (*Women Scientists in America*

[Baltimore: Johns Hopkins University Press, 1982], 16). This is possibly due to an onerous workload once she took the position at Barnard. Maltby herself was unable to find time for research after arriving at Barnard.

19 H. Brooks to L. Gill, 6 August 1906, BCA.
20 H. Brooks to L. Gill, 1 September 1906, BCA.
21 E. Rutherford to O. Hahn, 20 August 1906, Rutherford Collection of Correspondence.
22 M. Maltby to L. Gill, 1 September 1906, BCA.
23 L. Gill to H. Brooks, 6 September 1906, BCA.
24 H. Brooks to L. Gill, 13 September 1906, BCA.

CHAPTER EIGHT

1 Letter from John Martin to Mr Hoffer, 18 August 1947, from the archives, Essex County Historical Society, Elizabethtown, New York.
2 Robert J.A. Irwin is thanked for providing this information from the manuscript of his presentation to the Pundit Club, "Ad Majorem Amicitiae Gloriam," Buffalo, New York, 1987.
3 John Spargo, "With Maxim Gorky in the Adirondacks," *Craftsman* 11 (November 1906): 149–55.
4 Steve Barnett, "The Summer Maksim Gorky, Russian Writer, Spent at Keene," *Adirondack Daily Enterprise*, 9 September 1958, 1–2.
5 M.L. Porter, "East Hill: Two Experiments in Social Living," *North Country Life*, Fall 1957, 27–32.
6 Fabianism was a leftish middle-class movement that was popular among British intellectuals. See Norman MacKenzie and Jeanne MacKenzie, *The Fabians* (New York: Simon & Schuster, 1977).
7 Unidentified obituary entitled "Dr. John Martin Dead; Educator and Lecturer Had Camp at Keene", archives of the Essex County Historical Society, Elizabethtown, New York. Also see letter from Martin to Hoffer.
8 Prestonia Mann Martin, *Prohibiting Poverty*, 13th ed. (Winter Park, Fla.: National Livelihood Plan, 1939).
9 Prestonia Mann Martin, *(The Most Important Question in the World:) Is Mankind Advancing?* (New York: Baker and Taylor, 1910).
10 Prestonia Mann Martin, *Riding Lessons on Pegasus* (Winter Park, Fla., 193?).

11 John Martin and Prestonia Mann Martin, *Feminism, Its Fallacies and Follies* (New York: Dodd, Mead & Co., 1916).
12 Thomas Hale, "Maxim Gorky at Summerbrook" (Transcript of a talk and slide presentation given at the Keene Valley Library, 30 July 1979). We thank Dorothy W. Irving, librarian for the archives, Keene Valley Library, for a copy of this presentation.
13 We thank Robert J.A. Irwin for the rough notes he made from a tape of a talk given by John Martin "a number of years ago" at Keene Valley.
14 There are a number of translations of each Russian name. For example, we have used the form Andreyeva throughout, while other sources translate the name as Andreieva, Andreeva, Andre'eva, or Andrieva. Similarly, Gorky is sometimes found written as Gorki, Gorkey, or Gor'kij.
15 Jay Oliva, "Maxim Gorky Discovers America," *New York Historical Society Quarterly* 51 (January 1967): 45–60.
16 V.I. Lenin and A.M. Gorky, *Letters, Reminiscences, Articles* (Moscow: Progress Publishers, 1973), 259.
17 Alexander Kaun, *Maxim Gorky and His Russia*, reprint (New York: Benjamin Blom, 1968), 573.
18 Filia Holtzman, "A Mission That Failed: Gor'kij in America," *Slavic and East European Journal* 6 (1962): 227–35.
19 M.R. Werner, "L'Affaire Gorky," *New Yorker* 25 (30 April 1949): 62–73.
20 Jon Swan, "Innocents at Home," *American Heritage* 16 (February 1965): 58–61, 97–101.
21 Kaun, *Maxim Gorky and His Russia*, 386.
22 *World* (New York), 6 May 1906, 6.
23 *New York American and Journal*, 6 May 1906.
24 We thank L.P. Bykovtseva of the Gorky Museum for her research on our behalf. We are grateful to Michael Newton, Grenfell College, Corner Brook, for the translation of her notes.
25 Maria F. Andreeva, *Perepiska Vospominaniia* (Moscow: Iskusstvo, 1968), 391. We thank Steve Esh, University of California at Santa Cruz, for the translations.
26 Hale, "Maxim Gorky at Summerbrook."
27 *Sun* (New York), 22 June 1906, 11. The nearest station to Summerbrook was at Westport, where the train was met by Jim Heald, a former owner of the land on which Summerbrook stood. With his

team of farm horses, it was about a five-hour ride up to Hurricane Ridge (Hale, "Maxim Gorky at Summerbrook").

28 Nikolai E. Burenin, *Pamiatnye Gody: Vospominaniia* (Leningrad: Lenizdat, 1961). We thank Steve Esh, University of California at Santa Cruz, for the translations.

29 Notes from Bykovtseva, Gorky Museum.

30 We thank Edgar Jewett III for the opportunity to visit Summerbrook, tour the estate, and see the slides copied from the historical archives collected by Thomas Hale.

31 Burenin, *Pamiatnye Gody*.

32 *Archives of A.M. Gorky*, vol. 12 (Moscow, 1968), 187. We thank L.P. Bykovtseva of the Gorky Museum for the information, and Steve Esh, University of California at Santa Cruz, for the translation.

33 Burenin, *Pamiatnye Gody*.

34 Spargo, "With Maxim Gorky in the Adirondacks," 149–55.

35 Zina was his son, Zinovy Peshkov.

36 We believe the English translation of this letter to be in error, reading "a physics professor, and Miss Brooks, a nice old lady."

37 Maxim Gorky, *Letters* (Moscow: Progress Publishers, 1966), 49.

38 Notes from Bykovtseva, Gorky Museum.

39 Daniil S. Danin, *Rezerford* (Moscow: Molodaya gvardiya, 1967), 292. Danin notes: "Unfortunately, almost nothing is known about the details of the meeting of Rutherford and Gorky. In all, several lines were devoted to it in the unpublished memoirs of M.F. Andreyeva." We thank Steve Esh, University of California at Santa Cruz, for the translation.

40 We thank Lawrence Badash, University of California at Santa Barbara, for this information.

41 *New York Times*, 13 October 1906, 7.

42 *Sun* (New York), 14 October 1906, 14.

CHAPTER NINE

1 Nikolai E. Burenin, *Pamiatnye Gody: Vospominaniia* (Leningrad: Lenizdat, 1961). We thank Steve Esh, University of California at Santa Cruz, for the translation.

2 We thank L.P. Bykovtseva, Gorky Museum, for her research on our behalf. We are grateful to Michael Newton, Sir Wilfred Grenfell College, Corner Brook, for the translation of her notes.

3 L. Bykovtseva, "Gorkii in Italii," *Znamia* 44, no. 3 (1974): 178–209. We thank Dennis Bartels, Sir Wilfred Grenfell College, Corner Brook, for the translation of the relevant portions of the article.
4 *Avanti*, 27 October 1906.
5 *Avanti*, 28 October 1906; also quoted in Bykovtseva, "Gorkii in Italii," 181.
6 Burenin, *Pamiatnye Gody*.
7 Burenin, *Pamiatnye Gody*.
8 Burenin, *Pamiatnye Gody*.
9 *Avanti*, 1 November 1906.
10 Burenin, *Pamiatnye Gody*.
11 Henri Troyat, *Gorky* (New York: Crown, 1989), 110.
12 In the files of Harriet Brooks is a postcard – "Refugees Viewing the Approaching Fire, San Francisco" – sent on 15 November 1906 by Zinovy Peshkov from Oakland, California, to Mr Ivan Ladyshnikoff (for Harriet), 145 Uhlandstrasse, Berlin, Germany. The margin of the illustration contains the message "Goodby 'babushka'! – 'babushka' is sailing today. Far, far away! – Goodby! Zina." It is difficult to explain why Peshkov would send a postcard to Brooks via Ladyzhnikov. The message is also puzzling, since Brooks sailed to Europe in October.
13 Franklin Brooks III to Paul Brooks Pitcher, 22 October 1979. We thank Paul Brooks Pitcher for a copy of this letter.
14 Cited in George B. Kauffman, ed., *Frederick Soddy (1877–1956)* (Dortrecht: D. Reidel, 1986), 130.
15 Maria F. Andreeva, *Perepiska Vospominaniia* (Moscow: Iskusstvo, 1968). We thank Steve Esh, University of California at Santa Cruz, for the translation of this letter.
16 We thank Monique Bordry, Institut Curie, Paris, for searching the records and finding this entry.
17 André Debierne, born in Paris in 1874, was only sixteen years old when he started work with Pierre and Marie Curie. In 1899 he discovered the element actinium. After the death of Pierre Curie, he shouldered much of the work of the Curie laboratory, allowing Marie Curie to obtain the fame. He died near the Institut du Radium in 1949. His contributions are little known outside of France. See Gaston Dupuy, "Notice sur la vie et les travaux de André Debierne (1874–1949)," *Bulletin de la Société Chimique de France*, 1950, 1023–6.

18 Eve Curie, *Madam Curie* (New York: Doubleday, 1937).

19 Ellen Gleditsch, "Discours de Mme E. Gleditsch," in *Cinquantenaire du premier cours de la Marie Curie à la Sorbonne* (Cahors: A. Coueslant, 1957), 36. After her research with Curie, Gleditsch worked with Bertram Boltwood at Yale for a year, then returned to her native Norway and became a professor of chemistry at the University of Oslo. For details of her life, see Marelene F. Rayner-Canham and Geoffrey W. Rayner-Canham, "Pioneer Women in Nuclear Science," *American Journal of Physics* 58 (1990): 1036–43.

20 Lawrence Badash, "Decay of a Radioactive Halo," *ISIS* 66 (1975): 566–8. Pflaum notes that Curie was "glacial" and "haughty" except with her research workers, "to whom she gave unstintingly of herself" and whom she regarded as "her second family." See Rosalynd Pflaum, *Grand Obsession: Madame Curie and Her World* (New York: Doubleday, 1989), 153.

21 May Sybil Leslie to Professor Smithalls, 30 November 1909, Smithalls Collection, University of Leeds Library. We thank P.S. Morrish, sub-librarian, for supplying copies of the letters.

22 May Sybil Leslie to Professor Smithalls, 8 June 1911.

23 André Debierne, "Sur le dépôt de la radioactivité induite du radium," *Le Radium* 6 (1909): 97–106.

24 André Debierne, "Sur le coefficient de diffusion dans l'air de l'émination de l'actinium," *Le Radium* 4 (1907): 213–18.

25 L. Blanquies, "Comparaison entre les rayons α produits par différentes substances radioactives," *Le Radium* 6 (1909): 230–2.

26 Sir Arthur Schuster, born in Frankfurt, Germany, in 1851, received his undergraduate education at Owens College (later the University of Manchester). He obtained his doctorate at Heidelberg University and then worked as a researcher at the Cavendish Laboratory, Cambridge, before obtaining a post at the University of Manchester. His major discovery was that electricity is conveyed through gases by means of ions. When he resigned his chair at Manchester, he was instrumental in obtaining Rutherford as his successor. Schuster died in England in 1934.

27 Ernest Rutherford to Arthur Schuster, 25 March 1907, Royal Society Archives, London. N.H. Robinson, librarian, is thanked for supplying a copy of the letter and enclosed recommendation.

28 We thank Peter McNiven, archivist, John Rylands University Library of Manchester, for information on the Harling Fellowship.

29 Max Egelnick, librarian, Marx Memorial Library, London, is thanked for this information. This was the historic meeting at which the definitive split between the Bolsheviks and the Mencheviks occurred; see Alexander Kaun, *Maxim Gorky and His Russia* (New York: Jonathan Cape, 1931), 391.
30 Notes from Bykovtseva, Gorky Museum.
31 Andreeva, *Perepiska Vospominaniia.*
32 F. Brooks III to Paul Brooks Pitcher, 22 October 1979.
33 Scott Jamieson, French Department, Sir Wilfred Grenfell College, Corner Brook, is thanked for finding this letter with the assistance of Monique Bordry, archivist, in the archives of the Curie Institute. We also thank Cicily Grinling for her attempts to track down additional correspondence.
34 Montague Cohen, "My Dear Eve ... The Letters of Ernest Rutherford to Arthur Eve, 1907–1908," *Fontanus* 1 (1988): 3–37.
35 Ernest Rutherford to Arthur Schuster, undated, Royal Society Archives, London.

CHAPTER TEN

1 John Martin and Prestonia Mann Martin, *Feminism: Its Fallacies and Follies* (New York: Dodd, Mead and Co., 1917), 73.
2 Katherine R. Sopka et al., *Making Contributions: An Historical Overview of Women's Role in Physics* (College Park, Md.: American Association of Physics Teachers, 1984), 13.
3 Quoted in Roberta Frankfort, *Collegiate Women* (New York: New York University Press, 1977), 33.
4 The information on Frank Pitcher was compiled from *Who's Who in Canada (1912)* and from his obituary in *Gazette* (Montreal), 22 August 1935, 5. There are discrepancies between these two biographies, and we have chosen the information that is supported by other sources.
5 F.H. Pitcher and H.M. Tory, *A Manual Of Laboratory Physics* (New York: Wiley, 1901).
6 F.H. Pitcher, "The Effects of Temperature and of Circular Magnetization on Longitudinally Magnetized Iron Wire" (Papers from the Department of Physics, McGill University, no. 9, reprinted from the *Philosophical Magazine*, May 1899), 421–433.
7 We thank Paul Brooks Pitcher for the loan of these letters, referred

to below as from the PBP collection.
8 Frank Pitcher to Harriet Brooks, 31 December 1906, PBP.
9 Frank Pitcher to Harriet Brooks, 15 February 1907, PBP.
10 Frank Pitcher to Harriet Brooks, 22 April 1907, PBP.
11 We thank Shirley C. Spragge, assistant archivist, Queen's University, for searching the university records for 1907.
12 Frank Pitcher to Harriet Brooks, 22 April 1907, PBP.
13 Frank Pitcher to Harriet Brooks, 16 May 1907, PBP.
14 Frank Pitcher to Harriet Brooks, 1 June 1907, PBP.
15 Frank Pitcher to Harriet Brooks, 15 June 1907, PBP.
16 Frank Pitcher to Harriet Brooks, 15 February 1907, PBP.
17 Frank Pitcher to Harriet Brooks, 19 June 1907, PBP.
18 Frank Pitcher to Harriet Brooks, 21 June 1907, PBP.
19 Frank Pitcher to Harriet Brooks, 22 June 1907, PBP.
20 Frank Pitcher to Harriet Brooks, 27 June 1907, PBP.
21 Frank Pitcher to Harriet Brooks, 30 June 1907, PBP.
22 Frank Pitcher to Harriet Brooks, 2 July 1907, PBP.
23 Frank Pitcher to Harriet Brooks, 5 July 1907, PBP.
24 A. Stewart Eve to Ernest Rutherford, 7 or 8 July 1907, Rutherford Collection of Correspondence.
25 We thank Paul Brooks Pitcher for a copy of the marriage certificate.

CHAPTER ELEVEN

1 We thank Marijean A.S. Hodgson, Women's Canadian Club of Montreal, for the records of Brooks's addresses.
2 A. Stewart Eve to Ernest Rutherford, 24 March 1908, Rutherford Collection of Correspondence. We thank A.E.B. Owen, keeper of manuscripts, Cambridge University Library, for assistance in obtaining copies of this correspondence.
3 Elizabeth Fee, "Critiques of Modern Science: The Relationships of Feminism to Other Radical Epistemologies," in Ruth Bleier, ed., *Feminist Approaches to Science* (New York: Pergamon Press, 1986), 45.
4 Marilyn Bailey Ogilvie, "Marital Collaboration: An Approach to Science," in Pnina G. Abir-Am and Dorinda Outram, ed., *Uneasy Careers and Intimate Lives* (New Brunswick: Rutgers University Press, 1987), 104. Apart from the obvious example of the Curies, Winifred Beilby worked with her spouse, Frederick Soddy. See Marelene F.

Rayner-Canham and Geoffrey W. Rayner-Canham, "Pioneer Women in Nuclear Science," *American Journal of Physics* 58 (1990): 1036–43.

5 Lewis Pyenson, "The Incomplete Transmission of a European Image: Physics at Greater Buenos Aires and Montreal, 1890–1920," *American Philosophical Society (Proceedings)* 122 (1978): 92–114.

6 Mary Rutherford to Harriet (Brooks) Pitcher, 26 October 1907, Paul Brooks Pitcher Collection (PBP). We thank Paul Brooks Pitcher for the loan of this correspondence.

7 Mary Rutherford to Harriet (Brooks) Pitcher, 28 November 1907, PBP.

8 David Wilson, *Rutherford* (Cambridge, Mass.: MIT Press, 1983).

9 Mary Rutherford to Harriet (Brooks) Pitcher, 28 December 1907, PBP.

10 Mary Rutherford to Harriet (Brooks) Pitcher, 20 August 1908, PBP.

11 Mary Rutherford to Harriet (Brooks) Pitcher, 27 August 1908, PBP.

12 L.P. Bykovtseva, director of the Gorky Museum, Moscow, is thanked for information on Maria Andreyeva and Nikolai Burenin.

13 Prestonia Mann Martin to Harriet (Brooks) Pitcher, 28 October 1908, PBP. Unfortunately, none of the correspondence from Brooks to Prestonia Martin is in the Martin files at Rollins College, Winter Park, Florida. We thank Kathleen J. Reich, associate professor and archivist, for this information.

14 Prestonia Mann Martin to Harriet (Brooks) Pitcher, 4 November 1908, PBP.

15 The entry for Burenin in *The Great Soviet Encyclopedia* (vol. 4 [New York: Macmillan, 1973], 186), describes his exciting and dangerous activities "He [Nikolai Evgen'evich Burenin] completed responsible assignments for the Bolshevik Party, including missions in which he forwarded illegal Social Democratic literature and weapons from abroad to Russia, managed underground printshops and literature warehouses, arranged secret addresses, and obtained funds for Party purposes."

16 The entry for Andreyeva in *The Great Soviet Encyclopedia* (vol. 2, 91) notes that Andreyeva was persecuted by the police for her activity in the party. She later carried out Lenin's instructions to deliver the Bolshevik newspaper *Proletarii*, to collect and distribute materials on the history of the Russian Revolution, and to raise money and attract a wide circle of writers for the newspaper *Pravda*. After the

revolution, she had a number of theatrical successes and was subsequently promoted to director of the Moscow House of Scholars.

17 Prestonia Mann Martin to Harriet (Brooks) Pitcher, 4 November 1908.

18 We thank Paul Brooks Pitcher for this information.

19 Maria Andreeva, *Peripiska Vospominaniia* (Moscow: Iskusstvo, 1968), 156.

20 Notes from Bykovtseva, Gorky Museum.

21 Barbara M. Solomon, *In the Company of Educated Women* (New Haven, Conn.: Yale University Press, 1985), 122.

22 *Minutes of the McGill Alumnae Society 1901–17*. Phebe Chartrand, archivist, McGill University, is thanked for finding and obtaining copies of references to Brooks in the McGill Alumnae Archives.

23 Sonya (Sophia) Kovalevsky (1850–91) was an outstanding Russian mathematician. A detailed discussion of her life is given in Ann Hibner Koblitz, *A Convergence of Lives: Sofia Kovalevskaia, Scientist, Writer, Revolutionary* (Cambridge, Mass.: Birkhäuser, 1983).

24 Charlotte A. Scott (1858–1931) was a British-born mathematician. After an outstanding performance at the University of Cambridge, she set up the mathematics program at Bryn Mawr. Most of her thirty papers dealt with algebraic geometry. See Marilyn B. Ogilvie, *Women in Science: Antiquity through the Nineteenth Century* (Cambridge, Mass.: MIT Press, 1986), 158.

25 Mary Somerville (1780–1872) was an exceptional scientist of the early nineteenth century. See Elizabeth C. Patterson, *Mary Somerville and the Cultivation of Science, 1815–1840* (Boston: Kluwer, 1985); and Mary Somerville, *Personal Recollections*, reprint (New York: AMS Press, 1975).

26 Paul Brooks Pitcher is thanked for lending the complete manuscript of this presentation, which was included among Brooks's memorabilia.

27 For information on Ayrton, see Evelyn Sharp, *Hertha Ayrton 1854–1923* (London: Edward Arnold, 1926); on Maltby, see Edward T. Janes, ed., *Notable American Women 1607–1950*, vol. 3 (Cambridge, Mass.: Harvard University Press, 1971), 487; and on Sidgwick, see Ethel Sidgwick, *Mrs Henry Sidgwick* (London: Sidgwick and Jackson, 1938).

28 McGill Alumnae Society, *Annual Report 1914–15* (McGill University Archives).

29 McGill Alumnae Society, *Annual Report 1922–23* (McGill University Archives).
30 Catherine H. Joyce, *The First Forty Years of the University Women's Club of Montreal 1927–1967* (Montreal, 1962).
31 *Encyclopedia Canadiana*, vol. 2 (Toronto: Grolier of Canada, 1975), 188.
32 Alison Prentice et al., *Canadian Women: A History* (Toronto: Harcourt Brace Jovanovich, 1988), 203.
33 We thank Marijean A.S. Hodgson, Women's Canadian Club of Montreal, for researching the organization's files.
34 *Gazette* (Montreal), 12 April 1923, 3.
35 *Maple Leaf*, December 1923, 33.
36 As Brooks had attended plays and operas during her stay in Italy, this speaker would have held a particular interest for her.
37 The details of the meetings were obtained from the records of the Women's Canadian Club of Montreal.
38 Virginia C. Gildersleeve, *Many a Good Crusade* (New York: Macmillan, 1954).
39 *Gazette*, 5 April 1924, 5.
40 *Gazette*, 18 April 1933.

CHAPTER TWELVE

1 It is of note that three was the number of children recommended by her friend, Prestonia Martin. See Thomas Hale, "Maxim Gorky at Summerbrook" (Transcript of a Talk and slide presentation given at the Keene Valley Library, 30 July 1979). We thank Dorothy W. Irving, librarian for the archives, Keene Valley Library, for a copy of this presentation.
2 A. Stewart Eve, *Rutherford* (Cambridge: Cambridge University Press, 1939), 231.
3 Ernest Rutherford to Bertram Boltwood, 22 December 1914, in Lawrence Badash, ed., *Rutherford and Boltwood* (New Haven, Conn.: Yale University Press, 1969), 300.
4 We are greatly indebted to Paul Brooks Pitcher for his commentary and other assistance with this work.
5 We thank Cicely Grinling for this account.
6 We thank Joan Denny for this account.
7 *Gazette* (Montreal), 23 March 1929, 7.

8 We thank Franklin Brooks for this information.
9 *Gazette* (Montreal), 8 May 1929, 4.
10 *Gazette*, 25 March 1929, 6.
11 *Gazette*, 27 March 1929, 4.
12 *Gazette*, 28 March 1929, 13.
13 *Gazette*, 29 March 1929, 3.
14 *Gazette*, 8 May 1929, 4.
15 *Gazette*, 9 May 1929, 4.
16 Eve Curie, *Madame Curie* (New York: Doubleday, 1937), 198.
17 We thank Montague Cohen, McGill University, for this information.
18 Gordon F. Hull, "The New Spirit in American Physics," *American Journal of Physics* 11 (February 1943): 23–30.
19 William W. Nazaroff and Anthony V. Nero, *Radon and Its Decay Products in Indoor Air* (New York: Wiley-Interscience, 1988).
20 We thank Montague Cohen for copies of the letter from Rutherford to Eve, for the noted information on Rutherford's handwriting, and for a copy of the obituary from *Nature*.
21 Ernest Rutherford, "Obituary: Harriet Brooks (Mrs. Frank Pitcher)," *Nature* 131 (1933): 865.
22 *Gazette*, 18 April 1933. We thank Robert H. Michel, archivist, McGill University, for supplying this item.
23 *McGill News* 14, no. 3 (June 1933): 36. We thank Robert H. Michel, archivist, McGill University, for supplying this item.
24 *Gazette*, 22 August 1935.

CHAPTER THIRTEEN

1 Londa Schiebinger, "The History and Philosophy of Women in Science," *Signs* 12 (1987): 305–32.
2 Marianne Gosztonyi Ainley, "Introduction," in M.G. Ainley, ed., *Despite the Odds* (Montreal: Véhicule Press, 1990), 18.
3 P.L. Kapitza, "Recollections of Lord Rutherford," *Royal Society* (Great Britain), *Proceedings*, series A, A294 (1966): 123–37.
4 Carolyn Merchant, "Isis' Consciousness Raised," *ISIS* 73 (1982): 398–409. A typical example is Henry A. Boorse, Lloyd Motz, and Jefferson H. Weaver, *The Atomic Scientists: A Biographical History* (New York: Wiley, 1989), in which all the discoveries are tied to the famous names and no mention is made of the "supporting cast." The sections on Rutherford make no mention of Brooks.

5 These three communicated frequently, as is evident from Lawrence Badash, *Rutherford Correspondence Catalog* (New York: Center for History of Physics, 1974). The interaction between the major players is a classic example of the "invisible college phenomenon," where knowledge on a topic is first transferred by correspondence between the leading figures in the field. See Diana Crane, *Invisible Colleges* (Chicago: University of Chicago Press, 1972); and Daryl E. Chubin, *Sociology of Sciences: An Annotated Bibliography on Invisible Colleges, 1972–1981* (New York: Garland, 1983).
6 Ainley, *Despite the Odds*, 20.
7 Marianne G. Ainley, "Last in the Field? Canadian Women Natural Scientists, 1815–1965," in Ainley, *Despite the Odds*, 31.
8 Maurice Schofield, "Women in the History of Science," *Contemporary Review* 210 (1967): 204–6.
9 Mary E. Weeks, "The Discovery of the Elements. XIX. The Radioactive Elements," *Journal of Chemical Education* 10 (1933): 79–90; reprinted in M.E. Weeks and H.M. Leicester, *Discovery of the Elements* (Easton, Pa.: Journal of Chemical Education, 1968).
10 Robert K. Merton, "The Matthew Effect in Science, II," *ISIS* 79 (1988): 606–23. The Matthew Effect was named after the saying in the New Testament, the Gospel according to Matthew "For unto everyone that hath shall be given, and he shall have abundance; but from him that hath not shall be taken away even that which he hath."
11 In addition, many women considered their contributions to be too insignificant to entitle them to submit an entry.
12 Marilyn B. Ogilvie, *Women in Science: Antiquity through the Nineteenth Century* (Cambridge, Mass.: MIT Press, 1986); Patricia J. Siegel and Kay T. Finley, *Women in the Scientific Search: An American Bio-bibliography 1724–1979* (Metuchen, NJ: Scarecrow Press, 1985); and Margaret Alic, *Hypatia's Heritage: A History of Women in Science from Antiquity through the Nineteenth Century* (Boston: Beacon Press, 1986).
13 Caroline L. Herzenberg, *Women Scientists from Antiquity to the Present: An Index* (West Cornwall, Conn.: Locust Hill Press, 1986).
14 *The Macmillan Dictionary of Canadian Biography*, 4th ed. (Toronto: Macmillan of Canada, 1978); and *The Canadian Encyclopedia*, 2d ed. (Edmonton: Hurtig Publishers, 1988). We thank Montague Cohen, McGill University, for pointing out this fact.

15 Henry James Morgan, ed., *The Canadian Men and Women of the Time* (Toronto: William Briggs, 1912), 147. The entry refers to her as an "educationalist" and she is noted as "lately teaching in the US."
16 She has now been mentioned in a high school chemistry text: G.W. Rayner-Canham, P. Fisher, P. LeCouteur, and R. Raap, *Chemistry: A Second Course* (Toronto: Addison-Wesley, 1989), 566.
17 Lois Arnold, "Marie Curie Was Great, But ...," *School Science and Mathematics* 75 (1975): 577–84.
18 Goldie Morgentaler, "McGill Alumnae through the Decades," *McGill News* 65, no. 1 (Fall 1984): 20.
19 Margaret Gillett, *We Walked Very Warily* (Montreal: Eden Press Women's Publications, 1981).
20 Margaret W. Rossiter, *Women Scientists in America: Struggles and Strategies to 1940* (Baltimore: Johns Hopkins University Press, 1982).
21 Marelene F. Rayner-Canham and Geoffrey W. Rayner-Canham, "Pioneer Women in Nuclear Science," *American Journal of Physics* 58 (1990): 1036–43. This list does not include the names of women who worked in the field but never published their findings.
22 Margaret W. Rossiter, "'Women's Work' in Science, 1880–1910," *ISIS* 71 (1980): 381–98.
23 Vivian Gornick, *Women in Science* (New York: Simon & Schuster, 1983), 15.
24 Deborah J. Warner, "Women Astronomers," *Natural History* 88 (1979): 12–26.
25 Lawrence Badash, *Kapitza, Rutherford, and the Kremlin* (New Haven, Conn.: Yale University Press, 1985), 12.
26 John Lankford and Rickey L. Slavings, "Gender and Science: Women in American Astronomy, 1859–1940," *Physics Today* 43 (1990): 58–65. A prime example of the difficulties faced by women astronomers is described in Cecilia Payne-Gaposchkin, *Cecilia Payne-Gaposchkin: An Autobiography and Other Recollections* (Cambridge: Cambridge University Press, 1984).
27 J. Walkley and A. Gilchrist, "Canadian Content in First Year Chemistry," *Canadian Chemical News* 42 (1990): 29–30.
28 Nicholas Wade, "Discovery of Pulsars: A Graduate Student's Story," *Science* 189 (1975): 358–64. Bell notes that when her discovery was announced, reporters were more interested in knowing how her height compared to Princess Margaret's and how many

boyfriends she had. See George Greenstein, "Neutron Stars and the Discovery of Pulsars, Part Two," *Mercury* 14 (1985): 66–70.

29 Marie Curie, *Pierre Curie* (New York: Macmillan, 1923), 197.

30 Eugene Garfield, "Premature Discovery or Delayed Recognition – Why?" in *Essays of an Information Scientist*, vol. 4 (Philadelphia: ISI Press, 1981), 488–93.

31 Dolby argues the converse case: that work publicized by the scientific elite is more visible and more readily to be judged in terms of quality. See R.G.A. Dolby, "The Transmission of Science," *History of Science* 15 (1977): 1–43.

32 Joseph Spradley, "The Role of Women in Element and Fission Discoveries," *Physics Teacher* 27 (December 1989): 656–62.

33 Ainley, *Despite the Odds*.

34 Brush refers to the "Marie Curie Syndrome" – that is, young women students assume that success in science requires the slavish devotion that Curie showed. See Stephen G. Brush, "Women in Physical Science: From Drudges to Discoverers," *Physics Teacher* 23 (January 1985): 11–19.

BIBLIOGRAPHY

WOMEN IN SCIENCE

Abir-Am, Pnina G. "Essay Review: How Scientists View Their Heroes: Some Remarks on the Mechanism of Myth Construction." *Journal of the History of Biology* 15 (1982): 281–315.

Abir-Am, Pnina G., and Dorinda Outram, eds. *Uneasy Careers and Intimate Lives.* New Brunswick: Rutgers University Press, 1987.

Ainley, Marianne G. "D'assistantes anonymes à chercheures scientifiques: Une retrospective sur la place des femmes en sciences." *Cahiers de recherche sociologique* 4 (April 1986): 55–71.

– "Women Scientists in Canada: The Need for Documentation." *Resources for Feminist Research* 15 (November 1986): 7–8.

– , ed. *Despite the Odds.* Montreal: Véhicule Press, 1989.

Aldrich, Michele L. "Review Essay: Women in Science." *Signs* 4 (Autumn 1978): 126–35.

Alic, Margaret. *Hypatia's Heritage: A History of Women in Science from Antiquity through the Nineteenth Century.* Boston: Beacon Press, 1986.

Arena, Francesco. "Présence des femmes en science et technologie au Québec." *Cahiers de recherche sociologique* 4 (April 1986): 33–53.

Arnold, Lois. "Marie Curie Was Great, But ..." *School Science and Mathematics* 75 (1975): 577–84.

Bachtold, Louise M., and Emmy E. Werner. "Personality Characteristics of Women Scientists." *Psychological Reports* 31 (1972): 391–6.

Baker, Dale R. "Can the Difference between Male and Female Science Majors Account for the Low Number of Women at the Doctoral Level in Science?" *Journal of College Science Teaching* 13 (1983–84): 102–7.

Balka, Ellen. "Calculus and Coffee Cups: Learning Science on Your Own." *Resources for Feminist Research* 15 (November 1986): 11–12.

Becker, Lydia E. "On the Study of Science by Women." *Contemporary Review* 10 (1869): 386–404.

Birenbaum, Rhonda. "Against All Odds: Women in the Nuclear Industry." *Ascent*, Fall 1989, 21–8.

Blackstone, Tessa, and Helen Weinrich-Haste. "Why Are There So Few Women Scientists and Engineers?" *New Society* 51 (21 February 1980): 383–5.

Bleier, Ruth, ed. *Feminist Approaches to Science*. New York: Pergamon Press, 1986.

Brickhouse, Nancy W., Carolyn S. Carter, and Kathryn C. Scantlebury. "Women and Chemistry: Shifting the Equilibrium toward Success." *Journal of Chemical Education* 67 (February 1990): 116–18.

Briscoe, Anne M., and Sheila M. Pfafflin, eds. *Expanding the Role of Women in the Sciences*. New York: New York Academy of Sciences, 1979.

Brush, Stephen G. "Women in Physical Science: From Drudges to Discoverers." *Physics Teacher* 23 (January 1985): 11–19.

Chang, Hilda Lei, ed. *Proceedings of the First National Conference for Canadian Women in Science and Technology*. Vancouver: SCWIST, 1983.

Clay, R.W. "The Academic Achievement of Undergraduate Women in Physics." *Physics Education* 17 (1982): 232–4.

Cole, Jonathan R. *Fair Science*. New York: Free Press, 1979.

Cole, Jonathan R., and Harriet Zuckerman. "Marriage, Motherhood and Research Performance in Science." *Scientific American* 256 (February 1987): 119–25.

Cook, Gayl. "Women in Physics – Why So Few?" *American Journal of Physics* 57 (August 1989): 679.

Dresselhaus, Mildred S. "Women Graduate Students." *Physics Today* 39 (June 1986): 74–5.

Eisner, R. "Science's Future: Do Women Hold the Key?" *Scientist* 4, no. 20 (15 October 1990): 1, 10–12.

Erickson, Gaalen, Lynda Erickson, and Sharon Haggerty. *Gender and Mathematics/Science Education in Elementary and Secondary Schools*. Discussion paper 08/80, Ministry of Education, Province of British Columbia, 1980.

Ferguson, Janet, ed. *Who Turns the Wheel?* Ottawa: Science Council of Canada, 1981.

Ferry, Georgina, and Jane More. "True Confessions of Women in Science." *New Scientist* 95 (1982): 27–30.

Garfield, Eugene. "Why Aren't There More Women in Science?" In Eugene Garfield, ed., *Essays of an Information Scientist*, 498–505. Vol. 5. Philadelphia: ISI Press, 1983.

Gillbert, Catherine. "Making College Chemistry More Female-Friendly." *Canadian Chemical News* 41, no. 3 (1989): 18–19.

Gillett, Margaret. "The Majority Minority: Women in Canadian Universities." *Canadian and International Education* 7, no. 1 (June 1978): 42–50.

Glazer, Penina M., and Miriam Slater. *Unequal Colleagues: The Entrance of Women into the Professions, 1890–1940*. New Brunswick: Rutgers University Press, 1986.

Gold, Karen. "Get Thee to a Laboratory." *New Scientist*, 14 April 1990, 42–6.

Gornick, Vivian. "That Moment When You Suddenly See Something New in the World Is Like Nothing Else." *Ms* 12 (October 1983): 51–2, 130–5.

– *Women in Science*. New York: Simon & Schuster, 1983.

Grove, J.W. "Nonsense and Good Sense about Women in Science." *Minerva* 27 (1989): 535–46.

Grinstein, Louise S. "Women in Physics and Astronomy: A Selected Bibliography." *School Science and Mathematics* 80 (1980): 384–98.

Grissom, Abigail. "Top 10 Women Scientists of the '80s: Making a Difference." *Scientist* 4, no. 20 (15 October 1990): 18, 21.

Haas, Violet B., and Carolyn C. Perrucci, eds. *Women in Scientific and Engineering Professions*. Ann Arbor: University of Michigan Press, 1984.

Haggerty, Sharon M. "Secondary Science Education in Canada: Participation of Girls in Elective Science Courses." *Chem 13 News*, February 1991, 8–10.

"Handling Harassment." *Physics World*, January 1990, 12.

Harding, Jan. *Switched Off: The Science Education of Girls*. York, UK: Longman – Schools Council Publications, 1983.

Harding, Sandra. *The Science Question in Feminism*. Ithaca, NY: Cornell University Press, 1986.

Harding, Sandra, and Jean F. O'Barr, eds. *Sex and Scientific Enquiry*. Chicago: University of Chicago Press, 1987.

Hayes, Ellen. "Women and Scientific Research." *Science* 32 (1910): 864–6.
Herzenberg, Caroline L. *Women Scientists from Antiquity to the Present: An Index.* West Cornwall, Conn.: Locust Hill Press, 1986.
– "Women in Science during Antiquity and the Middle Ages." *Journal of College Science Teaching* 17 (November 1987): 124–7.
Herzenberg, Caroline L., Susan V. Meschel, and James A. Altena. "Women Scientists and Physicians of Antiquity and the Middle Ages." *Journal of Chemical Education* 68 (1991): 101–5.
Høyrup, Else. *Women of Science, Technology, and Medicine: A Bibliography.* Roskilde, Denmark: Roskilde University Library, 1987.
Hyde, Ida H. "Before Women Were Human Beings ..." *American Association of University Women, Journal* 31 (1938): 226–36.
Jansen, Sue Curry. "Is Science a Man? New Feminist Epistemologies and Reconstructions of Knowledge." *Theory and Society* 19 (1990): 235–46.
Kahle, Jane Butler. "Women Biologists: A View and a Vision." *BioScience* 35 (1985): 230–4.
– "Recruitment and Retention of Women in College Science Majors." *Journal of College Science Teaching* 17 (March/April 1988): 382–4.
– , ed. *Women in Science: A Report from the Field.* Philadelphia: Falmer Press, 1985.
Kass-Simon, G., and Patricia Farnes, eds. *Women of Science: Righting the Record.* Bloomington: Indiana University Press, 1990.
Kay, Marcia. "Girls = Science." *Canadian Living,* April 1991, 118–25.
Keller, Evelyn Fox. "Women in Science." *Harvard Magazine* 77 (1974): 14–19.
– *Reflections on Gender and Science.* New Haven, Conn.: Yale University Press, 1984.
– "Women and Basic Research: Respecting the Unexpected" *Technology Review* 87 (November/December 1984): 45–7.
– "Women Scientists and Feminist Critics of Science." *Daedalus* 117, no. 4 (1987): 77–91.
– "Feminist Perspectives on Science Studies." *Science, Technology, & Human Values* 13 (1988): 235–49.
– "Long Live the Differences between Men and Women Scientists." *Scientist* 4, no. 20 (15 October 1990): 15, 17.
Kelly, Alison. "Women in Science: A Bibliographic Review." *Durham Research Review* 36 (Spring 1976): 1092–1108.

– "Why Girls Don't Do Science." *New Scientist* 94 (20 May 1982): 497–500.
– ed. *Science for Girls?* Milton Keynes, UK: Open University Press, 1987.
Kistiakowsky, Vera. "Women in Physics: Unnecessary, Injurious and Out of Place?" *Physics Today* 33 (1980): 32–40.
Kohlstedt, Sally G. "In from the Periphery: American Women in Science, 1830–1880." *Signs* 4 (1978): 81–96.
Köppel, Anna-Pia. "Frauen in der Naturwissenschaft. Vom Mittelalter bis in die Gegenwart." *Feministische Studien* 4 (May 1985): 107–29.
Lankford, John, and Rickey L. Slavings. "Gender and Science: Women in American Astronomy, 1859–1940." *Physics Today* 43 (March 1990): 58–65.
Lewis, Ian. "Some Issues Arising from an Examination of Women's Experience of University Physics." *European Journal of Science Education* 5 (1983): 185–93.
Lonsdale, Kathleen. "Women in Science: Reminiscences and Reflections." *Impact of Science on Society* 20 (January–March 1970): 45–59.
Lloyd, Barbara. "Rules of the Gender Game." *New Scientist* 124 (2 December 1989): 60–4.
Lubkin, Gloria B. "Women in Physics." *Physics Today* 24 (April 1971): 23–7.
Luchins, Edith H. "Sex Differences in Mathematics: How NOT to Deal with Them." *American Mathematical Monthly* 86 (March 1979): 161–8.
McBay, Shirley M. "Inspiring Women to Pursue Science: A Job That Should Begin at Home." *Scientist* 4, no. 20 (15 October 1990): 15, 17.
MacLeod, Roy, and Russell Moseley. "Fathers and Daughters: Reflections on Women, Science and Victorian Cambridge." *History of Education* 8 (1979): 321–33.
McMillan, Ann. "Where Have All the Young Women Gone?" *University of Waterloo Alumni Magazine*, July 1990, 10–14.
Manthorpe, Catherine. "Feminists Look at Science." *New Scientist* 105 (7 March 1985): 29–31.
Matyas, Marsha Lakes. "Keeping Girls on the Science Track." *Curriculum Review*, January/February 1985, 75–8.
Megaw, W.J. "Girls' Physics Workshops at York." *Physics in Canada*, September 1986, 108.
Merchant, Carolyn. "Isis' Consciousness Raised." *ISIS* 73 (1982): 398–409.

Meyer, Gerald D. *Science for Englishwomen: 1650–1760.* Ann Arbor, Mich.: University Microfilms International, 1984.

Mitroff, Ian I., Theodore Jacob, and Eileen Trauth Moore. "On the Shoulders of the Spouses of Scientists." *Social Studies of Science* 7 (1977): 303–27.

Mozans, H.J. *Woman in Science.* Reprint. Cambridge, Mass.: MIT Press, 1974.

Murphy, Patricia. "Gender Gap in the National Curriculum." *Physics World,* January 1990, 11.

Naiman, Adeline. "Keeping the Computer neuter." *Personal Computing,* December 1985, 37.

Neuschatz, Michael. "Reaching the Critical Mass in High School Physics." *Physics Today* 42 (August 1989): 30–6.

Ogilvie, M.B. *Women in Science.* Cambridge, Mass.: MIT Press, 1986.

Phillips, Patricia. "Science and the Ladies of Fashion." *New Scientist* 95 (12 August 1982): 416–18.

– *The Scientific Lady: A Social History of Women's Scientific Interests 1520–1918* (London: Weidenfeld & Nicholson, 1990.

Pinet, Janine. "La femme et la science." *Atlantis* 3 (Spring 1978): 96–115.

Rebière, A. *Les Femmes dans la science.* 2d ed. Paris: Librairie Nony, 1897.

Reyes, Laurie H. "Sexual Stereotyping in Mathematics: Beyond Textbooks." *Arithmetic Teacher* 26 (April 1979): 25–6.

Rip, Arie. "Keller on Science Studies, or Reflexivity Revisited." *Science, Technology, & Human Values* 13 (1988): 254–61.

Romer, Robert H. "Editorial: 958 Men, 93 Women – How Many Lise Meitners among Those 865?" *American Journal of Physics,* 56 (1988): 873–4.

Rose, Hilary. "Comment on Schiebinger's 'The History and Philosophy of Women in Science: A Review Essay." *Signs* 13 (1988): 377–81.

Rosser, Sue V. "Good Science: Can it Ever Be Gender Free?" *Women's Studies International Forum* 11 (1988): 13–19.

– *Female-Friendly Science.* New York: Pergamon Press, 1990.

Rossi, Alice S. "Women in Science: Why So Few?" *Science* 148 (1965): 1196–1202.

Rossiter, Margaret W. "Women Scientists in America before 1920." *American Scientist* 62 (1974): 312–23.

– "'Women's Work' in Science, 1880–1910." *ISIS* 71 (1980): 381–98.

– *Women Scientists in America.* Baltimore: Johns Hopkins University Press, 1982.
– "Women and the History of Scientific Communication." *Journal of Library History* 21 (1986): 39–59.
– "Sexual Segregation in the Sciences: Some Data and a Model." *Signs* 4 (1988): 146–51.
Rundnagel, Regine. "Frauen in Naturwissenschaft und Technik: Ein Literaturbericht." *Das Argument* 28 (1986): 74–85.
Sabin, Florence. "Women in Science" *Science,* 83 (1936): 24–6.
Schiebinger, Londa. "The History and Philosophy of Women in Science." *Signs* 12 (1987): 305–32.
– *The Mind Has No Sex?* Cambridge, Mass.: Harvard University Press, 1989.
Schofield, Maurice. "Women in the History of Science." *Contemporary Review* 210 (1967): 204–6.
Scott, Joan Pinner. "Science Subject Choice and Achievement of Females in Canadian High Schools." *International Journal of Women's Studies* 4 (1981): 348–61.
Shapley, Deborah. "Obstacles to Women in Science." *Impact of Science on Society* 25 (1975): 115–23.
Sheinen, Rose. "The Rearing of Women for Science, Engineering and Technology." *International Journal of Women's Studies* 4 (1981): 339–47.
– "Women As Scientists: Their Rights and Obligations." *Journal of Business Ethics* 8 (1989): 131–55.
Siegel, Patricia J., and Kay T. Finley. *Women in the Scientific Search: An American Bio-bibliography 1724–1979.* Metuchen, NJ: Scarecrow Press, 1985.
Skolnick, Joan, Carol Langbort, and Lucille Day. *How to Encourage Girls in Math and Science: Strategies for Parents and Educators.* Toronto: Prentice-Hall, 1982.
Smail, Barbara. *Girl-friendly Science: Avoiding Sex bias in the Curriculum.* York, UK: Longman – Schools Council Publications, 1984.
Sopka, Katherine R., et al. *Making Contributions: An Historical Overview of Women's Role in Physics.* College Park, Md.: American Association of Physics Teachers, 1984.
Steinberg, Jonathan. "Nice Girls Do Biology." *New Society* 63 (17 March 1983): 429–30.
Sutton, Christine. "A Role Model for Female Physicists." *New Scientist* 103 (13 September 1984): 53.

Talbot, Marion. "Eminence of Women in Science." *Science* 32 (1910): 866.

Tobias, Sheila. "Math Anxiety and Physics: Some thoughts on Learning 'Difficult' Subjects." *Physics Today* 38 (June 1985): 61–8.

Tosi, Lucia. "Women's Scientific Creativity." *Impact of Science on Society* 25 (1975): 105–14.

Traweek, Sharon. "High-Energy Physics: A Male Preserve." *Technology Review*, November/December 1984, 42–3.

Trescott, Martha Moore. *Dynamos and Virgins Revisited: Women and Technological Change in History*. Metuchen, NJ: Scarecrow Press, 1979.

Tuana, Nancy, ed. *Feminism and Science*. Bloomington: Indiana University Press, 1989.

Vetter, Betty M. "Women in the Natural Sciences." *Signs* 1 (1976): 713–20.

Walford, Geoffrey. "Parental Attitudes and Girls in Physical Science." *School Science Review* 64 (1982–83): 566–7.

Walton, Anne. "Attitudes to Women Scientists." *Chemistry in Britain* 21 (1985): 461–5.

Warner, Deborah Jean. "Women Astronomers." *Natural History* 88 (May 1979): 12–26.

White, Martha S. "Physiological and Social Barriers to Women in Science." *Science* 170 (1970): 413–16.

Widnall, Sheila E. "AAAS Presidential Lecture: Voices from the Pipeline." *Science* 241 (1988): 1740–45.

Wilson, Joan Hoff. "Dancing Dogs of the Colonial Period: Women Scientists." *Early American Literature* 7 (1972/73): 225–35.

Zuckerman, Harriet, and Jonathan R. Cole. "Women in American Science." *Minerva* 13 (1975): 82–102.

BIOGRAPHIES OF WOMEN NUCLEAR SCIENTISTS

Barr, E. Scott. "The Incredible Marie Curie and Her Family." *Physics Teacher* 2 (1964): 251–9.

Curie, Eve. *Madam Curie*. New York: Doubleday, 1937.

Giroud, Françoise. *Marie Curie – A Life*. New York: Holmes & Meier, 1986.

Johnson, Karen E. "Maria Goeppert Mayer: Atoms, Molecules and Nuclear Shells." *Physics Today* 39 (September 1986): 44–9.

Kauffman, George B., and J.P. Adloff. "Marguerite Perey and the Discovery of Francium." *Education in Chemistry* 26 (1989): 135–7.

Kraft, Fritz. "Lise Meitner: Her Life and Times – On the Centenary of the Great Scientist's Birth." *Angewandte Chemie (International Edition)* 17 (1978): 826–42.

Kronen, Torleiv, and Alexis Pappas. *Ellen Gleditsch*. Oslo: Aventura Vorlag, 1987.

Lubkin, Gloria. "Chien-Shiung Wu, the Charming First Lady of Experimental Physics." *Smithsonian* 1 (1971): 52–7.

McCann, Mary. "No Parity in Science, Being a Woman Nuclear Physicist." *Science for People*, no. 46 (1980), 9–11.

Meitner, Lise. "Lise Meitner Looks Back." *Advancement of Science* 21 (1964): 39–46.

Opfell, Olga S. *The Lady Laureates*. Metuchen, NJ: Scarecrow Press, 1978.

Pflaum, Rosalynd. *Grand Obsession – Madame Curie and Her World*. New York: Doubleday, 1989.

Rayner-Canham, Geoffrey W., and Marelene F. Rayner-Canham. "The Shell Model of the Nucleus." *Science Teacher* 54, no. 1 (1987): 18–21.

Rayner-Canham, Marelene F., and Geoffrey W. Rayner-Canham. "Pioneer Women in Nuclear Science." *American Journal of Physics* 58 (1990): 1036–43.

Reid, Robert. *Marie Curie*. New York: Saturday Review Press, 1974.

Sime, Ruth L. "The Discovery of Protactinium." *Journal of Chemical Education* 63 (1986): 653–7.

– "Lise Meitner and the Discovery of Fission." *Journal of Chemical Education* 66 (1989): 373–6.

Spradley, Joseph. "The Role of Women in Element and Fission Discoveries." *Physics Teacher* 27 (1989): 656–62.

INDEX